Social Laser

Social Laser

Andrei Khrennikov

CRC Press
Taylor & Francis Group
Boca Raton London New York

CRC Press is an imprint of the
Taylor & Francis Group an **informa** business

Social Laser

First published 2020 by Jenny Stanford Publishing Pte. Ltd.

Published 2023 by CRC Press
Taylor & Francis Group
6000 Broken Sound Parkway NW, Suite 300
Boca Raton, FL 33487-2742

ISBN 978-981-4800-83-9 (hbk)
ISBN 978-0-367-81720-6 (ebk)

Visit the Taylor & Francis Web site at
http://www.taylorandfrancis.com

and the CRC Press Web site at
http://www.crcpress.com

British Library Cataloguing-in-Publication Data
A catalogue record for this book is available from the British Library.

To my lovely Anja

Contents

Preface

The recent years were characterized by stormy social protests throughout the world. By using an analogy from cosmology, we can say that the collapse of the Soviet Union was the singularity point for the Social Big Bang. These protests have some commonalities, but at the same time their sociopolitical, psychological, and economic contexts differ essentially. An important class of such protests is labeled *color revolutions.* Such events are often characterized by brutal clashes and lead to civil wars. Other social forms of protests take place within democratic societies. We can mention Brexit and the election of Donald Trump as president of the USA. Although the latter social actions are not brutal, their internal and global sociopolitical impact is even higher than the impact of some color revolutions.

The analysis of these events in social and political literature is characterized by a huge diversity of opinions.[1] And this is not surprising, because typically each author emphasizes some concrete feature in context of a social protest. Such sociopolitical biases lead to a variety of "explanations" of the recent worldwide waves of massive coherent social actions.

In contrast to the authors mentioned, we are not looking for concrete sociopolitical (or economic, or financial, or historical) roots of protests. We want to find the general conditions leading to the generation of waves of social actions, and generally we ignore the concrete social, political, economic, financial, and historical

[1]See, for example, Tucker [226], Demes and Forbrig [64], Butora [47], Fisher and Bijelic [80], Krastev [188, 189], Fukuyama [99], Mason [198], Mitchell [200], Schmidt and Cohen [218], Travers [225], Bouchet [35], and Putin [213].

variables involved in an event. (This book is neither on sociology nor on politics.)

How can one model some process without going deeply into details, without trying to determine and quantify the complex variables involved in the process? (We remark that in sociopolitical processes such variables are so complex that typically it is meaningless to even try to determine them not only quantitatively but also qualitatively.) Those who know the basics of quantum mechanics and its foundational problems (or those who have at least heard about them) can immediately recognize that this problem of social variables is similar to the problem of hidden variables in quantum mechanics. In the latter, one cannot determine variables permitting a finer state representation than those given by a wave function (complex probability amplitude). However, in this book we are not interested in the impossibility of using hidden variables for quantum processes. This is a complicated problem characterized by a diversity of opinions and stormy debates [153]. We are fine by recognizing that quantum formalism provides an excellent description of observations matching very well with experiments. And we want to proceed in the same way by modeling sociopolitical processes: to model observational outputs of social perturbations, but without attempting to find the "hidden variables" behind them.

We remark that the sociopolitical perturbations under consideration are characterized by the *cascade dynamics* leading to the exponential amplification of coherent social actions. In quantum physics, such exponential and coherent amplification is the basic feature of the functioning of laser. ("Laser" is the acronym for light amplification by stimulated emission of radiation.) In this book we explore (following papers by Khrennikov [168, 169, 176] and Khrennikov et al. [181]) the theory of laser to model the aforementioned waves of social protests, from color revolutions to Brexit and Trump's election. We call such social processes *stimulated amplification of social actions* (SASA), but to keep closer to the analogy of physics we shall merely operate with the term *social laser.* Our main efforts are directed toward distilling laser theory from optics, from light (i.e., we shall struggle against the letter L in "laser"). Instead of following the standard operating with frequencies or wavelengths, we shall operate with *social energy.* In physics, energy and frequency are coupled by a simple formula. But in social and political sciences, the notion of frequency seems to be ambiguous and we keep to the energy representation.

The notion of the social energy is the main novel component of our quantum-like modeling. To justify the use of the social analogue of physical energy, we use the quantum-mechanical interpretation of energy, not as an internal feature of a system, but as an observable quantity. Thus, as for an electron, we cannot assign to a human the concrete value of social energy. There are mental states in the superposition of a few different values of social energy. However, by designing proper measurement procedures we can measure human energy.

An important comment should be made on the first letter in SASA. Typically "stimulated" is associated with a kind of external ignition of amplification of social actions. We remind that in physics "stimulation" means that under the influence of electromagnetic radiation, laser's gain medium composed of, say, atoms starts emitting photons carrying the same characteristics as photons in the field (coherence). Such cascade-generating photons need not come from an external source; they can be generated in the gain medium by the process of spontaneous emission. The power of such a field (photons' flow) increases exponentially with the aid of a *laser resonator.* In physics, this is typically an optical cavity.

Similarly, spontaneous stimulation of emission can start in a social gain medium which approached population inversion. The latter means the number of humans, social analogues of atoms (s-atoms), occupying the excited state is higher than the number of humans occupying the non-excited state. In the same way as in physics, social lasers have to be equipped with resonators, e.g., in the form of *echo chambers*, which are typically Internet based.

On the other hand, similarly to physics, "stimulation" of social lasing can also be generated by an external information field, a portion of information communications injected in a human gain medium with population inversion. Social resonators also play a crucial role in the generation of lasing.

The word "stimulated" refers to both the forms of ignition of lasing:

- by a portion of spontaneously emitted messages,
- by a portion of externally injected messages.

For social engineering, the latter way of ignition is preferable.

We now spend some time to describe the present situation in quantum theory and why this situation is really our favorite for the realization of our social lasing project. The recent *quantum information revolution* has impacted not only the development of quantum theory, experiment, and technology, but also quantum foundations. The most fundamental problems of quantum theory have been described from another angle. (See, e.g., proceedings of the Växjö conferences on quantum foundations, 2000–2019; also see, e.g., the proceedings of the early conferences [1–5, 132, 138, 139, 143, 147, 155].) For our project of social lasing, the main consequence of the quantum information revolution is that nowadays quantum physical systems are widely treated as just carriers of information. So, quantum physics is rapidly transformed into quantum information theory. The physical properties of carriers of quantum information are often completely ignored. One is interested merely in the features of information processing. In particular, physical space-time has practically disappeared from quantum (information) theory. (As a consequence, the majority of experts in quantum information theory know very little about real quantum physics.) Quantum scenarios are considered in finite dimensional state spaces. Linear algebra became the main mathematical tool instead of functional analysis, which had been playing a crucial role during the first 70 years of the development of quantum mechanics.

Nowadays, one is not so much interested in the physical properties of *transformers of quantum information*. And it is natural to search for other classes of information transformers that are different from genuine quantum physical systems, but that transform information in a quantum-like manner. The latter means that we can use the basic tools of quantum formalism to model a transformer's behavior. The most basic tool is quantum probabilistic formalism, the calculus of complex probability amplitudes, endowed with *Born's rule* coupling complex amplitudes with probabilities. Another powerful mathematical tool is the use of *tensor products* to model (information) behavior of compound systems—a group of transformers interacting with information fields. Such interactions are modeled with the aid of the *theory of open quantum systems.*

Surprisingly, to find quantum-like systems, we need not go deep into the microworld, or isolate systems from environments, or freeze them to absolute zero. Anybody has such a system inside herself, and this is her brain. The analogy between the behavior of genuine quantum physical systems such as electrons or photons and psychological behavior of humans has been discussed since the first days of quantum mechanics. We can mention, e.g., the Pauli–Jung correspondence or Whitehead's philosophy (see Chapter 1 (Introduction) for details). However, only in the present quantum information era has this analogy begun to be formalized on the basis of quantum theory. In principle, this process of formalization can lead to the development of the social counterpart of laser theory with wide applications to social engineering. For the moment, social laser theory provides merely a qualitative characterization of the basic features of human gain mediums and information fields generated by the mass media leading to successful social lasing. In section 7.3, we made the first steps toward quantitative predictions.

Nowadays quantum-like models are widely used in psychology, decision making, economics, finances, social and political science, game theory, and modeling of the behavior of macroscopic biological systems such as cells or proteins [9, 106, 154] (see section 1.3.2 for details and references). The theory of social lasing is one of the recent fruits of quantum-like modeling in application to psychosocial processes.

As is well known, the analogy with the physical laser played a fundamental role in establishing the basic principles of *synergetics* [108–110]. Haken [111] emphasized that laser can be considered as a *trailblazer of synergetics.* He also pointed to "analogies between behavior of a laser threshold and phase transition" and to "significance of the laser as a prototype of systems which produce spatial or temporal structures of self-organization" [111]. This viewpoint is very supportive for the present project on social lasing.

At the same time, we have to point to the crucial difference between the quantum information modeling of social lasing presented in this book and the use of the analogy with the physical laser to motivate development of the science about cooperation: synergetics. Haken's idea was to use this analogy to unify in one framework the huge variety of cooperation processes. In such a project the quantum information basis of the functioning of laser was not—and, moreover, should not be—emphasized. The majority of "laser-like" synergetic processes have no relation to the quantum structure of information processing in social systems. In particular, Haken [111] considered classical laser-like equations (see Eqs. 13.1–13.3, p. 312) to illustrate the mathematical analogy with self-organization processes in classical physics. Our aim is different. We want to formalize quantum information features of social, and more generally information, systems that can lead SASA.

We also remark that quantum-like modeling has no direct coupling to the quantum physical brain project: an attempt to reduce cognition and consciousness to genuine quantum physical processes in the brain [103, 204, 215, 229] (see also [25, 29, 230]). At the same time we do not criticize this project. We maintain a neutral position. In any event, the quantum-like model for information processing by humans and, in particular, SASA are not based on the assumption that quantum physics plays some role in human cognition. Moreover, the quantum-like representation of information by the brain can be constructed on the basis of classical electrochemical processes in neural networks and uncertainty in the generation of action potentials [180]. However, the latter is not a topic of this book.

This book is the first step toward social laser engineering. For the moment, we derive only the most fundamental constraints onto human gain media, the functioning of social resonators, and the information field, leading to the possibility of successful social lasing. The social laser project (as well as the physical laser project) is very complex, and there a lot needs to be done before we approach the level of theory, measurement, and experiment as in the physical project. We list the main problems that must be resolved:

(1) to develop a comprehensive theory of social resonators (analogues of optical cavities); in this book we present only a schematic description
(2) to continue analysis of constraints on a human gain medium responsible for social lasing
(3) to study the role of information overload in switching humans into the indistinguishability regime of information processing in which the content of communication is practically ignored
(4) to analyze the impact of elimination of social diversity (in culture, lifestyle, food, nationality, gender, education, science, and economics) to prepare a proper human gain medium
(5) to elaborate and test measurement procedures for social energy and temperature
(6) to continue the determination of the basic parameters of social lasing (see section 7.3)

At the same time it is clear that in academic laboratories experimenting in social lasing is practically impossible. However, this should not be considered as a big problem of the social laser project. Nowadays the intensity of social lasing in the international arena has approached a high level, and scientists can make the best of this great opportunity for testing their models.

On the one hand, this book is oriented to the scientists (as well as students) from humanities, especially social science, who are interested in the applications of the social laser model and generally quantum-like models. They can proceed through Chapters 1–4 and Chapter 7 (besides section 7.3). As was emphasized, the quantum information revolution minimized the genuine physical content of quantum theory. The mathematical content also was essentially simplified. In any event, the basics of quantum theory and its mathematical apparatus are given in Chapter 10. Mathematically educated experts in economics, finances, and decision making can find the quantum-like model of subjective utility useful (Chapter 12).

On the other hand, this book is oriented to physicists who are interested in the applications of quantum physics and thermodynamics in social science. Besides the aforementioned sections, they can proceed to Chapter 7.3 and Chapters 5, 6, and 8. These are based on sufficiently advanced physics and mathematics.

Chapter 9 can be interesting for some scientists from humanities and natural science. But to read it one should really like Freud's ideas on unconscious–conscious coupling. For us, the most interesting material of this chapter is related to Freud's ideas on transformations between physical and psychical energies. This section contains numerous citations from Freud and is rather difficult to read.

Chapter 11 contains a discussion on the possibility of using the subjective interpretation of quantum probabilities. It foundationally supports the model presented in Chapter 12.

Andrei Khrennikov
Växjö, London, and Moscow
2016–2019

Chapter 1

Introduction

This introductory chapter is very detailed and it aims to describe the general state of the art in applications of the mathematical apparatus and methodology of quantum theory to humanities. One the one hand, I want to convince experts in humanities that quantum theory can resolve the well-known problems, in particular, in decision theory. On the other hand, I want to convince physicists and especially those who work in quantum foundations and quantum information and probability that applications of quantum theory to humanities are not exotic. (Nowadays top-level experts, such as psychologists, work actively on quantum-like modeling.)

In principle, this introductory chapter can be skipped and the reader can jump directly to Chapter 3 on the basics of physical laser theory (the physicist can jump directly to Chapter 4).

1.1 Interplay of Psychology and Physics: Historical Overview

The present sharp separation of physics and psychology is only a peculiarity of the present moment. The 19th and the first part of 20th century were characterized by mutual influence of physical and

Social Laser
Andrei Khrennikov

ISBN 978-981-4800-83-9 (Hardcover), 978-0-367-81720-6 (eBook)
www.jennystanford.com

Figure 1.1 Niels Bohr.

psychological theories and the fruitful exchange of ideas between the brightest representatives from both sides. One of the best-known examples is the impact made by psychology on quantum mechanics, which resulted in borrowing the *principle of complementarity* [33] by Niels Bohr from William James's book [120]. It may be less known that, in turn, the idea of complementarity was elaborated by James under the influence of 19th-century studies in thermodynamics, which led him (as well as, later, Sigmund Freud [84, 85]) to the notion of *psychic energy*. Initially, complementarity in psychology was about complementarity of different representations of psychic energy [120]: splitting of consciousness (dissociative identity disorder), which is characterized by at least two distinct mental representations. These are analogues of the position and momentum representations in quantum mechanics. Often, one speaks about the wave–particle duality. In one context a physical system behaves as a wave and in another context as a particle. In the same way a human with multiple personality disorder can exhibit a few totally different ("incompatible") personalities.

Reviewing a variety of definitions from dictionaries and encyclopedias, we believe that we can safely state the following. *Physics* is

Figure 1.2 William James.

the science that deals with the properties of matter. *Psychology* is the science that deals with mental processes and behavior.

In accordance with the views of Rene Descartes, there are two basic types of substance, material and mental, and one is not reduced to the other. (It is also a Buddhist dogma that life comprises mind and matter.) Although during the last century physical reductionism captured the headlines in psychology, Descartes' ideology still penetrates the body of modern science. Naturally, physics and psychology are considered different fields of science, each with its specific theoretical and experimental methodologies. It seems that there is nothing, or very little, in common between them. Most students of physics would probably not like to spend their time studying psychology courses, and most psychology students dislike physics.

However, developments in physics and psychology are connected much more strongly than one can imagine. We can point to a few big names who contributed to establishing a connection between the two most fundamental sciences (one about nature and the other about the psyche: Hermann von Helmholtz, Sigmund Freud,

Gustav Theodor Fechner, William James, Niels Bohr, Carl Jung, Wolfgang Pauli, Albert Einstein, ...

Freud was strongly influenced by the works of von Helmholtz on thermodynamics and especially on the energy conservation law. Of course, when discussing this law we have to mention the works of Germain Henri Hess, James Prescott Joule, and Rudolf Clausius. But, for Freud, the influence of von Helmholtz's ideas was especially strong. He started his research in physiology under the supervision of Ernst Brucke, who previously worked with Hermann von Helmholtz. He noted similarities between thermodynamics and the human psyche and developed a kind of mental thermodynamics known as *psychodynamics* [84–86]. Freud actively used the notion of the *psychic energy* (libido) and the law of its conservation.

Primarily libido represents sexual energy. However, according to Freud, sexual energy is one of the forms of psychic energy, which can be transformed into other forms (see Chapter 9). In the first stage of his psychodynamical studies, Freud was influenced by the ideas of Fechner: considering physical facts related to human body and mental facts as the sides of one reality. Fechner concluded that both physical and mental phenomena must be described by the same mathematical apparatus. This remark is very important for us as foretelling the main idea of this book:

> *Both mind and matter behaviors nicely fit the framework of the mathematical formalism of quantum theory.*

The notion of psychic energy played an important role in theorizing by James [120]. Following physicists who at that time were already using field theory, he started to operate with the notion of *psychic field*. This psychic field as well as a physical field can have different modes. This analogy led James [120] to the fundamental *principle of complementary* of information belonging to different modes of consciousness:

> It must be admitted, therefore that *in certain persons*, at least, *the total possible consciousness may be split into parts which coexist but mutually ignore each other,* and share the object of knowledge between them. More remarkable still, they are *complementary.*

> Give an object to one of the consciousnesses, and by this very act you remove it from the other or others. Barring a certain common fund of information, like the command of language, etc., what the upper self knows the under self is ignorant of, and vice versa.

Now we point to the famous correspondence between Pauli and Jung [12, 122] on comparative analysis of the foundations of physics and psychology. These letters were written in a free style of discussion between friends (and, in part, a patient and a psychoanalyst). At the beginning Pauli wanted to discuss with Jung his psychical problems which might be a subject of psychoanalytic treatment. However, Jung smartly redirected Pauli to a young female psychoanalyst, and most of the Pauli–Jung correspondence is about psyche–physics interrelation. This freedom of exchange allowed them to express in psychoanalytic manner many thoughts which would be never presented in formal scientific discussions. From the letters it is clear that Jung was deeply influenced by quantum theory [12, 122] in Pauli's presentation; e.g., Jung wrote to Pauli [122]:

> As the phenomenal world is an aggregate of the processes of atomic magnitude, it is naturally of the greatest importance to find out whether, and if so how, the photons (shall we say) enable us to gain a definite knowledge of the reality underlying the mediative energy processes. Light and matter both behave like separate particles and also like waves. This . . . obliged us to abandon, on the plane of atomic magnitudes, a causal description of nature in the ordinary space-time system, and in its place to set up invisible fields of probability in multidimensional spaces.

Inspired by acausal features of quantum mechanics, Jung developed his famous theory of *synchronicity* [121], the theory about the experiences of two or more events as meaningfully related, where they are unlikely to be causally related. The subject sees it as a meaningful coincidence. The use by quantum physicists of "invisible fields of probability on multidimensional spaces" strongly supported Jung's interest in the psychic fields, invisible, probabilistic, and defined not on the physical space time $\mathbf{R}^4$, but on some kind of "mental space"; cf. with p-adic and more general treelike models of mental spaces developed by the author of this book [145]. This was a clue to unification of psychic and quantum physical fields in one *psychophysical field.* The idea was very appealing to both Pauli and Jung and it was one of the topics of their correspondence. Jung also discussed field models with Einstein, and Einstein's attempts to create a unified pure field model of physical reality (see, e.g., Einstein and Infeld [75]) also stimulated Jung's studies on psychical fields. However, neither the Einstein dream for the purely field description of physical reality nor the Jung–Pauli dream [122] for the unified psychophysical field were presented in the rigorous mathematical framework.

Our discussion on mutual influence of physics and psychology can be shortly represented as the following (of course, incomplete) diagram:

[Hess, Joule, Clausius, and von Helmholtz] $\Rightarrow$ [Freud, Fechner, James] $\Rightarrow$ [Bohr] $\Leftrightarrow$ [Pauli] $\Leftrightarrow$ [Jung] $\Leftrightarrow$ [Einstein] . . .

1.2 Quantum Brain

The "quantum brain" idea was very popular 20–30 years ago (see, e.g., [103, 204, 215, 229]; see also [25, 29, 30, 230]). These attempts to reduce cognition and even consciousness to genuine quantum physical processes in the brain are rooted in the strong belief that quantum theory is the ultimate physical theory. Believers of the ultimateness of quantum theory think that all processes in nature can be reduced to quantum physical processes. In particular, the brain's functioning and its "products," cognition and consciousness, are generated by such processes. We point to the three main streams in the quantum brain project:

(1) Quantum Mechanics: consciousness from entanglement in microtubulas (Hameroff [103]).
(2) Quantum Field Theory: consciousness as the states of the quantum field generated by the brain (started by Umezawa

[229], see also [215], and nowadays this QFT-brain model is actively developed by Vitiello [230, 231]).

(3) Quantum Gravity: consciousness as the chain of collapses of superpositions of masses in the brain (Penrose [204]).

These streams have their own advantages and disadvantages. We are not interested in the physical problems related to the quantum brain approach. For us it is crucial that all quantum brain models suffer from the impossibility to couple them to the paradigm of the neuronal processing of cognitive information. The neurons are not the basic entities of these models, but just some supplementary physical structures. And this attitude to diminish the role of neurons in cognition has a clear explanation.

All authors developing the genuine quantum physical models of the brain's functioning pointed to the impossibility of considering neurons in a state of superposition. For example, Vitiello [230] pointed out:

> "In the quantum model 'we do not intend,' Ricciardi and Umezawa [215] say, 'to consider necessarily the neurons as the fundamental units of the brain."

(Similar statements can be found in the works of Penrose [204], Hameroff [103], and Bernroider [30].) As a consequence, in such models neurons were not considered as the basic units of information processing. And this viewpoint on neurons' role in information processing diminished the interest of experts in neurophysiology and cognitive science to the quantum brain models. Since 1990 I participated in a few conferences such as "Towards Science of Consciousness" (Tuscon, Arizona). These were great conferences gathering the top experts and general public. They played the crucial role in stimulating my early steps towards the science of consciousness. I remember that neurophysiologists were really angry by hearing the exciting talks of Hameroff and Penrose.

We again stress that the topic of this book has nothing to do with the attempts to model cognition and consciousness from the genuine quantum physical processes in the brain.

1.3 Quantum-Like Modeling of Cognition and Decision Making

The author of this book was pioneering in establishing the quantum-like approach for modeling cognition and consciousness [130] and their applications to psychology and cognitive science [145, 154]. My first step towards this approach was based on the analysis of the probabilistic foundations of quantum theory. I tried to separate genuine physical components from the purely mathematical ones [131]. Generally this book is not about probability. And the reader may skip the next section and jump to section 1.3.2. On the other hand, the reader who is interested in the probabilistic structure of quantum theory can enjoy reading Chapter 10 (mathematical formalism of quantum theory), Chapter 11 (QBism, subjective probability interpretation), and Chapter 12 (quantum probabilistic model of subjective expected utility).

1.3.1 From Probabilistic Foundations of Quantum Mechanics to Quantum-Like Modeling

During 1980–1990 my life was characterized by a strong interest in the foundations of probability theory, especially alternatives to the Kolmogorov measure-theoretic approach [191]. (I recall that the modern axiomatics of classical probability theory is based on this approach.)

The crucial point was understanding that in the probabilistic framework *quantum interference*[1] can be interpreted as a deformation of the *formula of total probability* [133, 136, 141, 145, 146, 149, 153, 154]. The quantum analogue of the formula of total probability contains an additional summand, the interference term. The appearance of this term is a consequence of the context-dependence of probabilities [153]. My way to understanding the contextual structure of quantum experiments was not straightforward.

---

[1] As is shown, for example, in the two slit experiments with photons or electrons.

As was mentioned, I was interested in alternatives to the Kolmogorov theory and especially in von Mises's frequency probability theory [232–234]. In the latter, probability is directly defined as the limit of frequencies of realizations of possible outputs of an experiment. Any experimental context generates a sequence of outputs, a collective.[2] In Kolmogorov's measure-theoretical approach the approximation of probability by frequencies is based on a mathematical theorem, the law of large numbers.

I applied the von Mises theory to quantum experiments [131, 145, 152, 153]. Surprisingly, all quantum paradoxes (and all such paradoxes have a probabilistic nature) disappeared. The next step was understanding the contextual nature of the von Mises theory: each collective is generated by the concrete experimental context. Hence probabilities depend on generating collective (experimental context). We remark that von Mises even considered compatible-incompatible collectives: the probabilistic counterparts of compatible-incompatible observables. Finally, I understood that contextual probability theory can generate quantum-like effects such as interference of probabilities (violation of the formula of total probability [133, 136, 141, 145, 146, 149, 153, 154]) and violation of Bell-type inequalities [8, 9, 22, 57, 70–72, 145, 154]. The general formalism of contextual probability theory was presented in [153, 154].[3]

The quantum-like paradigm was formulated [130]:

> *The mathematical formalism of contextual probability, in particular, quantum probability, can be used to model not only the statistical behavior of genuine quantum physical systems, e.g., electrons, but all context-sensitive systems, e.g., humans. Contextual information processing cannot be based on complete resolution of ambiguity. Therefore such systems process ambiguities, process superpositions of alternatives.*

---

[2]The probability of the outputs labeled by α (e.g., "head" in the coin tossing experiment) is given by $\lim_{N\to\infty} n_\alpha/N = p_\alpha$, where n_α is the number of realizations of label α in the first N trials.

[3]I also made essential efforts to find the answer to another question: What are the constraints on contextual probabilities leading to the representation in the complex Hilbert space? This is a very complex mathematical problem [153, 154]. In the general case, the answer is not known.

The main distinguishing feature of quantum probability theory[4] is the *linearity* of its formalism. Quantum probability is probabilistic linear algebra. Linear models always attract by their simplicity. In particular, quantum probabilistic formalism is characterized by the linear representation of superpositions. (Generally, in contextual probability theory superpositions are represented nonlinearly through the formula of total probability with the interference term). Therefore it is always useful to try to proceed with quantum probability by modeling contextual behavior.

1.3.2 Quantum-Like Models Outside Physics

Nowadays quantum-like modeling is widely used in mathematical modeling in cognitive science, psychology, decision making, game theory, economics and finance, social and political sciences (see, e.g., the pioneering paper [130], monographs, [9, 14, 45, 106, 107, 145, 154] and a few representative papers [6, 8, 11, 15, 22, 23, 36–38, 44–46, 105, 140, 142, 161, 163, 166, 184, 187, 210, 211, 221]. These applications are typically based on the use of quantum probabilistic calculus, instead of the classical one.

It was demonstrated that experimental statistical data collected in cognitive psychology, decision making, social science, game theory, molecular biology, and epigenetics demonstrates quantum probabilistic features [9, 14, 45, 106, 107, 154]. For example, such data can violate the classical formula of total probability and this violation can be mathematically expressed in the form of quantum interference [9, 44–46, 106, 107, 130, 145, 154, 210, 211]. Some cognitive experiments demonstrating the violation of Bell-type inequalities have been performed [8, 9, 22, 57, 70–72, 145, 154]. The violation can be interpreted as *contextuality of mental observables* represented as questions or tasks.

In theoretical studies, cognitive psychology, and decision making, quantum probability provides the adequate mathematical models for the basic psychological effects such as *conjunction, disjunction,*

[4]The probabilistic calculus based on the representation of probabilities by states, normalized vectors in complex Hilbert space, and observables by Hermitian operators or more generally by so-called positive operator valued measures.

and order effects. Its use also resolves the fundamental paradoxes of decision theory such as *Allais* [7], *Ellsberg* [76], and *Machina* [193] *paradoxes*. The resolution of the paradoxes of the classical decision theory (including expected utility, subjective utility, and prospect theories) and creation of the paradox-free (at least up to now) decision theory played a very important role in the justification of applications of quantum probability outside physics.

We also recall that as early as the 1970s, Tversky (one of the most cited psychologists of all time) and Kahneman (winner of the Nobel Prize in economics, in 2002, for prospect theory, which he co-developed with Tversky) have been demonstrating numerous cases where classical probabilistic prescription and human behavior persistently diverge [227, 228].[5] From their viewpoint, which is widely shared in the decision-making community, we are at a theoretical crossroads, with huge divisions across conflicting, entrenched theoretical positions.

- Should we continue relying on (classical) probability as the basis for descriptive and normative predictions in decision making (and perhaps ascribe inconsistencies to methodological idiosyncrasies)?
- Should we abandon probability theory completely and instead pursue explanations based on heuristics, as Tversky and Kahneman proposed?

However, the use of the probabilistic and statistical methods is really the cornerstone of modern scientific methodology. Thus, although the heuristic approach to decision making cannot be discarded completely, it seems more natural to search novel probabilistic models for decision making. Quantum probability model can be considered as a possible alternative to the heuristics of Tversky and Kahneman [227, 228] (see Chapter 12 for further details, discussions, and references).

---

[5]These studies were coupled to the problem of rationality of decision makers. In fact, in decision theory rationality is defined on the basis of classical probability theory. A rational decision maker is characterized by using the Bayesian update of probability. Therefore violations of rationality can be interpreted as violations of the laws of (classical) probability.

1.4 Operational Formalism: Creation and Annihilation Operators

We emphasize that the quantum formalism provides only a formal operational description of physical processes; in particular, the spontaneous emission and stimulated absorption and emission which play the fundamental role in lasing theory.

One of the important mathematical representations of the operational formalism is given in terms of *creation and annihilation operators* $\widehat{a}^{\star}$, $\widehat{a}$. In quantum optics,

- operator $\widehat{a}^{\star}$ represents creation of a photon through its emission;
- operator $\widehat{a}$ represents disappearance of a photon from the field resulting from absorption of this photon by an atom.

Here photons are considered *excitations of the quantum electromagnetic field.* Thus, creation and annihilation operators represent emission and absorption of excitations.

These are linear operators acting in complex Hilbert space representing the states of the quantum field. The operator $\hat{a}^{\star}$ is adjoint to the operator $\hat{a}$. The operator $\hat{n} = \hat{a}^{\star}\hat{a}$ plays an important role in quantum field theory. This is the operator of the photon number. Its eigenstates are states $|n\rangle$ of the quantum field with the fixed number n of photons. Similar operational description can be given for the processes of state transitions in atoms.

The operational formalism is very useful in quantum theory, since here one cannot construct a more detailed description in terms of classical-like variables, known as hidden variables (at least this is generally assumed; see section 4.6 for the discussion on this issue). Therefore one proceeds with operational formalism, representing preparations of systems and observations.

In this book, operational formalism including creation and annihilation operators is used to model the process of social lasing and stimulated amplification of social actions (cf. [14, 187]). Here a crucial role is played by the information field (the analogue of the electromagnetic field in physical lasing) and its excitations (the analogues of photons) carrying portions of social energy:

- operator $\hat{a}^*$ represents emission of an information excitation;
- operator $\hat{a}$ represents absorption of an information excitation.

One of the reasons for application of quantum operational formalism to modeling social processes is that, similarly to quantum physics, it is (practically) impossible to present a detailed account of all sociopsychological factors involved in stimulated amplification of social actions. In this situation, the operational description in terms of formal absorption and emission of information excitations can be fruitful.

We can refer to stormy discussions in sociopolitical literature (see section 2.2) on the origin of color revolutions. These discussions are characterized by diversity of opinions and the impossibility to present a detailed account of the interrelation between all social, psychological, political, and economical and financial factors leading to such protest waves. The operational quantum model of "social laser" is based on ignoring these factors (playing the role of hidden variables; see section 4.6). We are not interested in the difference in, e.g., the emotional states of people in different countries nor the difference in the political situations in these countries. The operational formalism formally describes the processes of absorption and emission of portions of social energy; see sections 1.7, 5.3 for further justification (from the information-theoretcial and thermodynamical principles) of applying the quantum formalism to the modeling of social lasing.

1.5 Social Laser as a Fruit of the Quantum Information Revolution

In this book we are interested in the information component of quantum theory. We recall that the recent quantum information revolution generated *the information interpretation of the quantum theory.* In fact, a few often competing information interpretations were proposed [39, 43, 48–51, 58, 59, 89, 97, 144, 149, 175, 205, 209, 238]. However, all of them are characterized by a paradigm shift: quantum systems are treated as merely carriers of information. Thus, from this viewpoint, quantum theory is about information flows. Hence, it may be possible to apply the quantum information approach even outside physics by identifying the quantum-like features of processes under consideration.

In [168, 169, 176] the author proposed the quantum-like model of the social (or more generally information) laser, the social analogue of the physical laser. Social laser theory was used for the mathematical description of the *stimulated amplification of social actions* [176]. In principle, social laser's functioning can be modeled without exploring the mathematical formalism of modern quantum mechanics based on the complex Hilbert space representation of states and observables. We can proceed by using thermodynamics in the same way as Einstein did by creating the theory of spontaneous and stimulated emission and absorption (see Chapter 6). However, the use of modern quantum formalism is more convenient for concrete technical calculations (see section 7.3).

The model of the social laser is based on the treatment of humans as carriers of *social energy*. In [168, 169, 176], the notion of social energy was (rather schematically) discussed in relation to the social lasing model. We shall present a detailed analysis of this notion based on the operational approach to quantum theory and its information interpretation [39, 43, 48–51, 58, 59, 89, 97, 138, 144, 149, 175, 205, 209, 238] (section 4.1).

In the simplest setting we consider a two-level cognitive system with relaxation and excitation states. We call such a system a (two-level) *social atom*, s-atom. In this book we are focused on considering two energy levels systems (see, however, section 6.3). In quantum-like modeling, the discrete structure of the social energy levels of s-atoms, energy quantization, has no straightforward relation to neurophysiology. In particular, the two-level structure is based on, so to say, "to be or not to be" scaling of social energy. This is a kind of emotional quantity. Of course, emotions are coupled to physical and neurophysiological processes in the human body, but this coupling is very complicated. And its concrete structure is not important for us.

A *social gain medium* is composed of s-atoms. They interact with the information field generated by the mass media and the Internet. This field is modeled as a quantum information field carrying quantized portions of social energy, social (or information) excitations. This field was formally introduced in [168, 169, 176]. In this book we put in essential efforts to propose the proper interpretation of the excitations of this quantum field as carriers of social energy (section 4.1.2).

In addition to social energy, we introduce the notion of the *social color* of an information excitation (section 4.3). This quantity represents additional characteristics of information excitations linked to the lasing coherence of the type of polarization and collimation (section 4.7).

1.6 Bose–Einstein Statistics of Information Excitations

For the non-physicist, we remark that quantum systems are either *bosons or fermions*. For example, photons are bosons and electrons are fermions.

In heuristic terms, bosons are systems which enjoy being together in the same state. Roughly speaking, if some boson S finds that in a room there are other bosons and in the same state ψ as S,

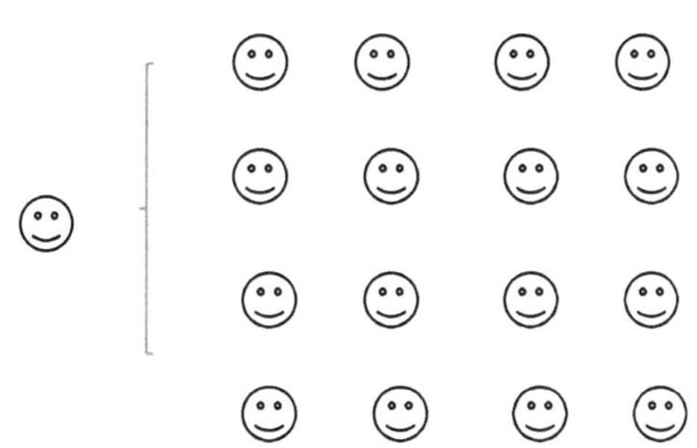

Figure 1.3 A boson-agent (the left-hand side) finds in the room (the right-hand side) a boson-crowd. This crowd is very attractive for this agent, who is really happy to join it. (All agents, including the newcomer, are smiling.) The probability that this agent would join the group is high. Guys in the crowd are indistinguishable from each other. This indistinguishability is face indistinguishability. Of course, the agents in the crowd can be distinguished by their locations in space. However, if agents move quickly, then it is practically impossible to identify the position of the agents with identical faces.

Figure 1.4 A boson-agent (the left-hand side) finds in the room (the right-hand side) just a few boson-agents. This group is not so much attractive for the newcomer (dull face). The probability that this agent would join the group is not so high. Guys in the group are (face-)indistinguishable from each other.

e.g., equally excited as S, then S strongly wishes to join this bosonic company. The probability that boson S would enter a room increases with the number of bosons (in the same state) which have already occupied this room (see section 4.8). If excited S enters a room occupied by one hundred excited bosons, S would like to join them. If S enters a room with ten thousand excited bosons, then his wish to join them essentially increases (see again section 4.8). We illustrate bosons' behavior by Figs. 1.3 and 1.4.

Fermion S likes to stay alone. If some fermion S sees in a room another fermion S' in the same state ψ as S, then S would never enter this room. However, if the states of S and S' are different, e.g., they have different degrees of excitement, say S is excited and S' is completely relaxed, then S joins S' with pleasure. The latter situation illustrates the behavior of electrons in an atom. An atom is a house with many rooms, electron orbits, each room corresponding to the concrete level of excitement. Each room, orbit, can be occupied by only one electron.[6] We illustrate fermions' behavior by Figs. 1.5 and 1.6.

---

[6]Well, although this picture represents the essence of fermionic behavior, it is oversimplified. The state ψ of a quantum system is determined by not only the level of excitement but also energy. An electron has another characteristic called spin. And an electron's state is determined not only by energy but also by spin. The same

Figure 1.5 A fermion-agent finds that some room is empty. This empty room attracts the agent (smiling face).

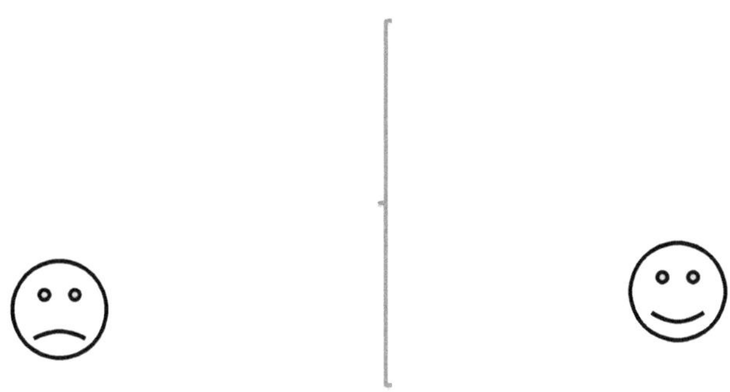

Figure 1.6 A fermion-agent (the left-hand side) finds in the room (the right-hand side) another fermion-agent. This room repels the newcomer (dull face).

Typically the Bose–Einstein (bosons) and Fermi–Dirac (fermions) statistical behaviors are associated with the specialty of quantum physical systems. (Classical systems follow the Maxwell–Boltzmann statistics; see below.) One says: "Oh! Photons and

---

room corresponding to the concrete energy level can be occupied by electrons with different spins.

electrons are really mystical creatures. It is not surprising that statistically they behave differently from the classical gas (e.g., air molecular in a room)." However, the thermodynamical analysis of the behavior of gases shows that the only distinguishing feature leading to quantum statistics is the *indistinguishability* of systems composing a gas (see section 5.3, where we follow Schrödinger's book [220]). Thus the only constraint is the indistinguishability of systems occupying the same state.

This thermodynamical analysis is applicable to ensembles of indistinguishable systems of any origin, i.e., not only genuine quantum physical systems. One may say that in any event this analysis is inapplicable to ensembles of humans: humans are definitely distinguishable! By replying to such an argument we now point to the essence of the notion of indistinguishability in the thermodynamical reasoning. This is indistinguishability with respect to some concrete observable A, i.e., not "complete indistinguishability," but A indistinguishability. Typically A is the energy observable $A = E$. Moreover, in accordance with the Copenhagen interpretation (which can be applied not only to quantum mechanics but also to any theory of measurement) the outputs of the A measurement are not objective properties of system S. These outputs are generated in the process of observation. This interpretation of indistinguishability can be applied to humans (see sections 4.1, 4.4).

The classical gas is composed of distinguishable systems; corresponding statistics is known as the *Maxwell–Boltzmann statistics.* This statistics reminds of the Bose–Einstein statistics. However, there is one important difference playing the crucial role in lasing. Consider the average number of systems $\langle n \rangle$ as a function of a system's energy

$$f(\epsilon) = \langle n \rangle(\epsilon).$$

By comparing the upper (Bose–Einstein) and intermediate (Maxwell–Boltzmann) graphs in Fig. 1.7, we can see that, for the same energy, the average number of bosons is essentially higher than the average number of classical systems. This feature of Bose–Einstein statistics is crucial for lasing. It is possible to create large ensembles of systems of sufficiently high energy.

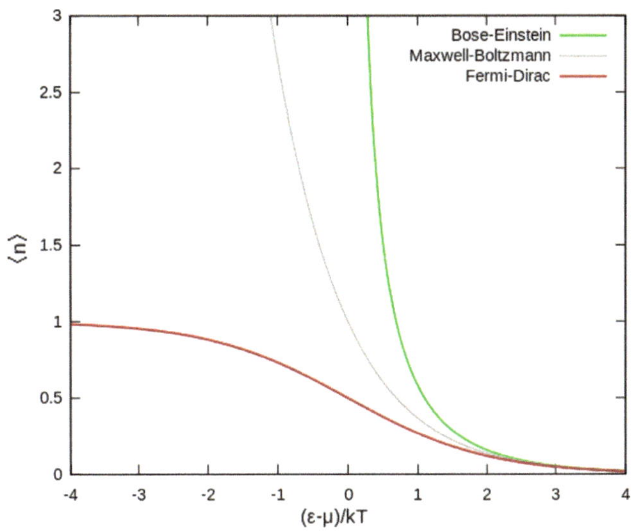

Figure 1.7 Graphs for Bose–Einstein, Maxwell–Boltzmann, and Fermi–Dirac statistics; dependence on a system's energy ϵ and temperature T (and so-called chemical potential μ).

Classical systems also like to spend time together, but they dislike too crowded rooms occupied by excited people. Their wish to join a fest decreases with an increase in excitement essentially quicker than in the case of bosons.

The role of (in)distinguishability can also be illustrated in heuristic terms. "Classical people" in the room are distinguishable. And a newcomer, say Alice, is confronted with a crowd of concrete personalities. This is not a simple situation, especially if these people are excited. Psychical pressure may be too strong. Therefore Alice may prefer to leave this crowded room immediately. "Quantum-like people" are not distinguishable. A newcomer does not care about their presence, say as at a railway station, subway, or airport. Of course, this is a naive illustration which should be taken too seriously.

In Fig. 1.7 we see that, for all statistics, function $f(\epsilon)$ decreases with an increase in energy. Differences between statistics slowly disappear. All statistics predict that only a few (passionate) guys can approach very high energies.

In physics, probability distributions also depend on another important parameter, temperature (see Fig. 1.7). In this book we shall deeply explore the notion of social energy. It is possible to introduce even a social analogue of temperature (see section 5.3.1 for a discussion).

1.7 Powerful Information Flows as the Basic Condition of Social Laser Functioning

The basic feature of quantum information fields is the indistinguishability of excitations carried by them. This is indistinguishability with respect to the social energy carried by excitations (see section 4.4 for further information).

People overloaded by information do not put essential efforts to analyze the contents of communication delivered the by mass media. They function as absorbers of social energy delivered by communication emitted by TV, newspapers, and the Internet. In particular, to approach population inversion, the mass media can pump into a human gain medium a flow of shock news about various catastrophes, violence, and political scandals. Agents absorb social energy. Later, this energy can be liberated and directed through the laserlike process of stimulated emission.

The power of information flows plays a crucial role in the creation of quantum-like processes of absorption and emission of social energy.

Nowadays these flows are extremely powerful. Hence, it is easier to create social (information) lasers than, say, it was 150–200 years ago.

1.8 Resonators of Physical and Social Lasers

In physics, the laser resonator in the form of an optical cavity plays a crucial role in the process of amplification of the output beam and makes it coherent. We model the functioning of social laser's resonator by "distilling" the physical scheme (see Chapter 7) to exclude the straightforward connection with optics. We proceed with the *quantum master equation for the density operator* describing excitations of the quantum information field inside the resonator of the social laser (see section 7.3.3). The main aim is establishing proper social interpretations of the basic quantities and parameters of this dynamical system.

The Internet-based *echo chamber* is considered an important example of social resonators. Its functioning is mathematically represented by the field of social excitations in the form of posts and comments interacting with the human gain medium.

One of the basic consequences of quantum-like dynamics is the existence of the threshold value for the pump parameter (see section 7.3.5): if the power of pumping is essentially higher than this threshold, then practically all social energy pumped into a human gain medium is transferred into the output beam of social actions.

This is a good place to emphasize once again the role of the huge power of the information flows generated by modern mass media.

Chapter 2

Social Laser Model for Stimulated Amplification of Social Actions

In recent years, we have seen the occurrence of a high level of social protests throughout the world, see, e.g., Krasev [188]:

> In the five short years between Occupy Wall Street and Vladimir Putin's "Occupy Crimea," we witnessed an explosion of protests all around the world—the Arab Spring, Russian Winter, Turkish Summer, and the dismembering of Ukraine all were part of the protest moment. Each of these demonstrations—and many less monumental ones—was angry in its own way, but the protests are also a worldwide phenomenon.

The problem of the "explanation" of the recent social explosions, especially in the Middle East, but also in Southern Europe, and the USA, has been debated actively in the social and political literature. We can mention the contributions of Tucker [226], Demes and Forbrig [64], Butora [47], Fisher and Bijelic [80], Krastev [188, 189], Fukuyama [99], Mason [198], Mitchell [200], Schmidt and Cohen [218], Travers [225], Bouchet [35], and Putin [213] to this debate. However, the diversity of opinions and conclusions is really amazing. At the moment, there is no consistent and commonly acceptable sociopolitical theory of these events.

Social Laser
Andrei Khrennikov

ISBN 978-981-4800-83-9 (Hardcover), 978-0-367-81720-6 (eBook)
www.jennystanford.com

2.1 What Can Be Expected from the Social Laser Model?

We propose to use the social laser model for such social events. In the last sentence the word "model" plays a crucial role. The social laser model is quantum-like. Such models are operational. They can represent formally process structure, but cannot "explain" its origin. In this way quantum-like models are similar to the genuine quantum model. The experience gained from quantum physics supports the use of such operational models, although they are not able to provide a kind of explanation of processes.

Reaction to application of such a methodology in social science may be negative. One may say that the social laser model ignores the most interesting questions such as

What are the basic social, economic, political, and historical causes for the Maidan revolution in Kiev or for the election of Donald Trump in the USA?

However, ignoring such problems does not devaluate quantum-like models. By using the social laser model it is possible to derive constraints on human population, "gain medium," and the information field generated by the mass media and the Internet, leading to stimulated amplification of social actions. In the same way as in physics, quantum-like models, including the social laser, have predictive power (see section 7.3). The main casualty of quantum-like social models is that their predictions are still qualitative. In contrast to physics, it is still impossible to present the concrete numerical values of the model's parameters guaranteeing initiation and successful functioning of stimulated amplification of social actions. The main problem is the absence of a well-developed methodology for measuring social energy.

In physics some researchers try to complete the operational quantum model by finer models having more explanatory power. The most well known example is Bohmian mechanics. In contrast to the operational quantum formalism, the latter describes the particles' trajectories. However, such subquantum models have merely a foundational value. They are essentially more complicated

Figure 2.1 Georgia's Rose Revolution (Tbilisi 2003).

than the operational quantum model. In principle, one can try to construct similar subquantum-like models for social processes. (See, e.g., the works [14, 52–54, 104, 106, 145] for applications of Bohmian mechanics in behavioral finances.)

Another subquantum model, prequantum classical statistical field theory (PCSFT), was proposed by the author [157–160, 162, 174]. In this model, quantum systems are represented by classical random fields. This a purely wave model of quantum phenomena. In contrast to the nonlinear mathematical structure of Bohmian mechanics, PCSFT is a linear theory. (See [156] for its application of modeling the brain's functioning.)

2.2 Color Revolutions

Here we consider a special form of social explosions known as *color revolutions.* Each of such revolutions is characterized by its special color label (often associated with some flower). Such coloring plays

Figure 2.2 Ukraine's Orange Revolution (Kiev 2004).

an important role (see section 4.3) in social lasing. Coloring is crucial for the reation of sharp output beams of social actions from human gain media.

The series of color revolutions (see map at Fig. 2.3)[1] started at territory of the former Soviet Union and Balkans, for example, Yugoslavia's Bulldozer Revolution (2000), Georgia's Rose Revolution (2003, Fig. 2.1),[2] Ukraine's Orange Revolution (2004, Fig. 2.2),[3] Kyrgyzstan's Tulip Revolution (2005), Belarus's Jeans Revolution (2006), and Moldova's Grape Revolution (2009). Further examples are presented by color revolutions outside of the former Soviet Union and Balkans, e.g., Lebanon's Cedar Revolution (2005, Fig. 2.4),[4] Kuwait's Blue Revolution (2005), and Myanmar's Saffron Revolution (2009). Some researchers trace the roots of color

[1]By A. Katsaris, https://commons.wikimedia.org/w/index.php?curid=3253293
[2]By Zaraza, https://commons.wikimedia.org/w/index.php?curid=1668494”> Link</a>
[3]By M. Duimel, https://commons.wikimedia.org/w/index.php?curid=9027926
[4]By E. Ghobeira, https://commons.wikimedia.org/w/index.php?curid=18961029

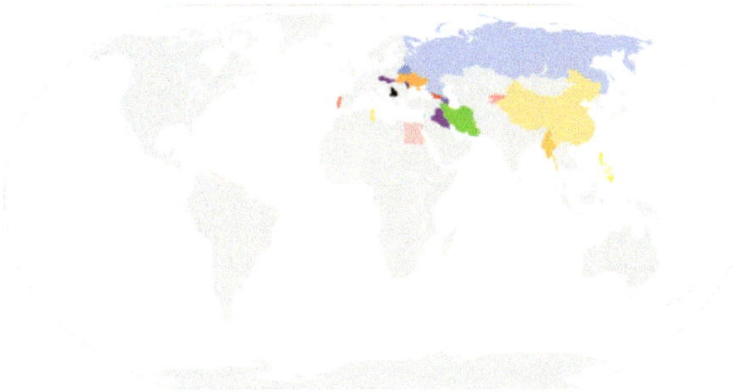

Figure 2.3 The map of color revolutions.

revolutions to the 1986 People Power Revolution (also known as the Yellow Revolution) in the Philippines. Recent years were characterized by a wave of color revolutions at the Middle East: Tunisia's Jasmine Revolution (2011), Egypt's Lotus Revolution (2011), and Muslim military uprisings in Iraq, Libya, and Syria. We can also mention Iran's Green Revolution (2009) and China's Jasmine Revolution (2011), Ukraine's second color revolution, the Maidan revolution (2014), and the following civil war in Ukraine.

These revolutions were the subject of numerous studies and publications in political and social science: see, e.g., Tucker [226], Demes and Forbrig [64], Butora [47], Fisher and Bijelic [80], Krastev [188, 189], Fukuyama [99], Mason [198], Mitchell [200], Schmidt and Cohen [218], Travers [225], Bouchet [35], and Putin [213]. The color revolutions are definitely the new sociopolitical phenomenon. They differ crucially from, e.g., the French Revolution or the Great October Revolution. However, by recognizing the specialty of color revolutions, different authors treat them in very different ways: from folk uprisings against corruption to a new form of warfare and terrorism. For the latter viewpoint, see, e.g., the citation of Putin [213]:

> In the modern world extremism is being used as a geopolitical instrument and for remaking spheres of influence. We see what tragic consequences the wave of so-called color revolutions led to.

Figure 2.4 Lebanon's Cedar Revolution (Beirut 2005).

The majority of authors point to common features of these sociopolitical events in spite of differences in interpretations:

- no clear programs, aims, and long-term plans;
- no real leaders compatible with the leaders of "real revolutions" such as Marx, Lenin, Trotsky and Stalin, Orozco, Villa and Zapata, Atatürk, Mussolini, Hitler, Mao, Castro, and Che Guevara;
- no ideology as was expressed in Marx's *Capital*, Lenin's *Imperialism, the Highest Stage of Capitalism*, Trotsky's *The Permanent Revolution*, or Hitler's *Mein Kampf*.
- "quick relaxation"—the explosion-like character of events.

2.3 Democratic Social Protests

Recent events that have happened inside the Western democratic system show similar features: the Greek bailout referendum (2015), Brexit (2016), election of Donald Trump as president of the USA

(2017). For example, Travers [225] presented the viewpoint that Brexit can be considered as antigovernment vote:

> The Conservatives have been in power since 2010, albeit as the leaders of the coalition from 2010–15. Facing an opposition which even its own (largely ex-) frontbenchers believe to be unelectable, it is small wonder some voters sought refuge in a referendum protest vote.

And he continues:

> 2008 and the banking crisis: The economic havoc caused by the 2008 banking crisis has never led to any obvious downside for many of those who caused it. Inflated boardroom pay has been sustained, with a massive gap in earnings between senior executives and shop-floor workers. Many on the centre-right of politics share with those on the left a view that something is wrong with this outcome. The public almost certainly believes it.

A similar viewpoint regarding Trump's election in the USA was presented in a series of interviews collected by Fishwick [81]. For example, he presented an interview with Rachael, 34, Indiana, small business owner:

> I want to change America to serve the people instead of a political system that wants to serve itself. My life won't change much, except I will have more hope that my government is trying to make our country strong instead of pandering to its own liberal interests. My main hope is that he will help balance our budget, and secondly that he will help our economy to remain strong.

2.4 Social Energy Pumping

It is easy to see patterns of social energy pumping preceding the aforementioned social protests. And this is also one of the basic components of functioning of social lasers.

Figure 2.5 Ukraine's Maidan revolution: massive protests in Kiev (2014).

Consider, for example, the Maidan revolution (see Figs. 2.5 and 2.6)[5] During a few years preceding it, the Ukrainian and international mass media generated high-intensity information flow of shocking news about the corruption of President Yanukovych and Ukrainian politicians in general.[6] This information flow was combined with the equally intensive flow of negative news about Russian policy and Russians in general. The important role in energy pumping was played by massive events at which big crowds of young people were charged with social energy through simple, but extremely energetic, slogans, such as "if you're not jumping, you're a Moskal," "one language, one nation, one homeland," and "hang the Russians."

---

[5] By A. Hikari, https://commons.wikimedia.org/w/index.php?curid=31272490; by M. Chernov, https://commons.wikimedia.org/w/index.php?curid=30763852

[6] This corruption news-flow started long before Yanukovych. The news about corruption in the apparatus and among relatives of Presidents Kuchma and Yushchenko were massively pumped by Ukrainian newspapers and TV and echoed in social network chambers.

Figure 2.6 Ukraine's Maidan revolution: brutal uprising in Kiev (2014).

This way of social energy pumping in human gain media was already tested during the previous massive protests on the postcommunist territory. In particular, we can mention one of the first energy-pumping campaigns, OK '98 (Slovakia, 1998)[7]:

> The largest and most visible OK '98 project was "Road for Slovakia," organized by the civic association GEMMA 93. During a 15-day march (August 19–September 3, 1998), some 350 civic activists covered more than 850 towns and villages across Slovakia, distributing 500,000 brochures to inform voters about the elections. Door-to-door campaigning explained voting procedures, stressed basic principles of parliamentary democracy and emphasized the importance of citizen participation in the

[7] "Slovak nongovernmental organizations earned considerable public reputation in the struggle against the nondemocratic practices of Prime Minister Vladimir Meciar and his government. With the parliamentary elections in September 1998, this struggle came to a head. A consolidated platform of NGOs launched the civic campaign known as OK '98 (full name: Obcianska kampan '98) to increase citizens' awareness about the elections, to encourage them to vote and to guarantee a fair ballot through independent civic supervision" (see Butora [47]).

> elections. Theatre performances featuring popular actors were an integral part of this activity (see Butora [47]).

The OK '98 project also pumped social energy in the Slovak gain medium by using simple slogans carrying a huge charge of social energy; e.g., see again paper [47]:

> Aiming to attract young voters, the civic association Hlava '98 (Head '98) organized a series of TV and radio spots airing the slogan "I think, therefore, I vote. I vote, therefore, I am" in cinemas, on the private station TV Markiza, on TV NASA in Eastern Slovakia and twelve private radio stations covering different areas of the country. In a non-traditional and artistically imaginative way, these spots stressed freedom and the importance of voting. In the ads, young people called on their peers to take part in the elections. Young people's role models, like athletes and actors, relayed the message. The ads were chic, genuine, humorous, and they had the charm of personal expression.

From the viewpoint of content, the Ukrainian slogans such as "one language, one nation, one homeland" and "hang the Russians" have nothing in common with the aforementioned Slovak slogan "I think, therefore I vote. I vote, therefore I am," but they are very similar from the viewpoint of social charging.

Already in Slovakia mass media–based delivery of social energy played a crucial role. Energy pumping is impossible without the involvement of financial flows, as was emphasized in the paper by Butora [47]:

> A large-scale campaign, such as OK 98, requires considerable resources and funding. An effective system and procedure for the submission of NGO projects was created by the Donors' Forum, an informal association of grantmaking foundations supporting democracy and civic participation. '32 Donors built a flexible funding system, simplified application procedures and provided co-financing for projects. Information about these activities was regularly published in the Non-Profit magazine and on the Internet. While the campaign benefited from an enormous amount of voluntary work, from in-kind contributions like, for instance,

> the creative input of artists and professionals specializing in "social campaigns," as well as from smaller contributions made by local business and private donations, the support provided by European and U.S. donors, coming from both the private and public sectors, was of critical importance. [These included the Civil Society Development Foundation, the Open Society Foundation, the Foundation for a Civil Society, the Children of Slovakia Foundation, the Carpathian Foundation, the Charles Stewart Mott Foundation, the German Marshall Fund of the United States, the Jan Hus Educational Foundation, the British Know How Fund, the Fund of Canada, the United States Information Service (USIS), and others. The overall financial volume of the campaign was estimated at approx. 30 million Slovak crowns or, at that time, US$857,000.]

In [80], we can find data on the funding of GLAS 2000 campaign for the presidential elections in Croatia, including international funds such as the Charles Stewart Mott Foundation, the British Know How Fund, the USAID Mission, and the U.S. Embassy's Democracy Commission.

Now consider such an unexpected event as voting for Brexit in the UK. In fact, the UK mass media, although positioned as generally liberal, worked properly to pump social energy to Brexit supporters. Emigration was the hot topic in the European mass media: boats with hundreds people approaching the European coast, people infiltrating the UK through the tunnel under the English Channel, huge crowds moving from South to North towards Germany, the UK, Sweden. In the UK mass media, local economic problems were coupled with emigration from East Europe, especially from Poland. Such topics as corruption and nonfunctionality of costly EU bureaucracy were presented and discussed in thousands of posts and videos which were spread via the Internet and social networks.

In the USA, the process of population inversion in the social laser with Trump as the output (Fig. 2.7)[8] started with Clintons' couple. Their sexual and financial affairs and widespread accusations in

[8]By White House photographer: Official White House Facebook page, Public Domain, https://commons.wikimedia.org/w/index.php?curid=55185771

Figure 2.7 President Donald Trump being sworn in on January 20, 2017, at the U.S. Capitol building in Washington.

corruption were massively highlighted by mass media and the Internet. This campaign played an important role in pumping social energy into the Trump-oriented gain medium. Obama's policy, especially his attempt of medical insurance reformation, can be considered as the continuation of the energy-pumping phase. The USA's mass media and social networks worked intensively to present failures of the functioning of American presidents and governments in both internal and foreign affairs, e.g., in Iraq, Iran, Syria, North Korea, Ukraine, and Russia. This energy pumping was crucial for population inversion: converting the initially neutral part of the American electorate into excited supporters of Trump's slogans.[9]

[9] Nowadays we can see new energy pumping generated by Trump's actions, but into another gain medium, oriented to the Democratic Party. But, in turn, the recent protest actions of liberals work as energy pumping in the Trump-oriented gain medium . . . The same happens in the UK. Brexit, as well as following actions of the British prime minister, Theresa May, pumped social energy into the labor-oriented gain medium.

2.5 Quick Relaxation

All aforementioned events are characterized by quick social relaxation. And this is one of the basic features of spike-like emissions generated by social lasers.

Consider, for example, the Maidan revolution in Kiev. Nowadays, in spite of a tremendous decrease in income and increase in the level of corruption, the Ukrainian population is being in the ground state, the state of complete social relaxation.

The same can be said about the situation in the USA after Trump's victory. The part of population which served as the "gain medium" for Trump's election is totally relaxed, although the new administration can propose only cheap populist actions.

In the UK, Brexit came as a shock wave. Now it is fully evident that the real Brexit has not so much to do with the expectations of voters. However, Brexit supporters are staying peacefully in the ground state of complete relaxation.

Perestroika in the USSR is the best example of rapid relaxation to the ground state of social energy. In 1990, in spite of having fallen to real poverty and destruction of industry and agriculture, science, education, and culture, the Russian population was staying in the ground state. All social energy was realized in the Perestroika spike, in the form of massive protest actions as in my own town, Zelenograd, near Moscow. We, the students and teachers of Moscow Institute of Electronic Engineering, radiated a huge amount of social energy in demonstrations and meetings against the corrupt communist bureaucracy in Uzbekistan, known as the Uzbek mafia.

This rapid relaxation is in striking contrast to situations caused by "real revolutions," such as the French Revolution and the Great October Revolution. Real revolutions are not followed by relaxation to the ground state of social energy. After the revolution, French people were so strongly excited that only the battle at Borodino and the Russian winter of 1812 were able to cool them down. In the same way, the post-revolution excitement of the population of Soviet Russia was really amazing.

2.6 Echo Chambers

We emphasize the crucial role of the *Internet echo chambers* in organizing protest actions nowadays (see section 7.2 for quantum-like modeling). Their functioning has led to an essential increase in the approachability of individuals, even those who distance themselves from absorption of news generated by the conventional mass media in the form TV, newspapers, and journals.

Anti-globalization actions were definitely the first tests of the possibility of pumping social energy into human gain mediums with the aid of the Internet echo chambers as well as of approaching the homogenization of the output spike of the corresponding social lasing. Anti-globalization social lasing started to function properly in Prague in September 2000 with a strong spike involving around 10,000 protesters. The next spike was in Quebec City in April 2001. The strongest spike was in Gothenburg in June 2001 with around 25,000 activists. Three activists were shot and injured. And then, one of the last strong spikes was in Genoa in July 2001, where one activist was shot dead by the police. As well as color revolutions, anti-globalization actions led to quick relaxation. Now these actions are practically forgotten.

Figure 2.8 Anti-globalization protests in Gothenburg, 2001.

2.7 Conflating Opposition Protests with Warfare

We have already pointed to the diversity of opinions on the origin of the color revolution. One of the extreme viewpoints is that these revolutions should be treated as warfare. The Russian president Vladimir Putin is one of the most prominent opponents of such an interpretation. The detailed analysis of the consequences of this viewpoint on color revolutions was presented in article [35]:

> The tendency of foreign actors to overstate their role in colour revolutions has declined but Western governments are still tempted to express support for ostensibly pro-democracy protesters, as happened in Ukraine in 2014. Western reactions to the anti-government protests in Armenia, Macedonia and Moldova last year were more low-key, partly because these were smaller and the countries less important, yet Russia's Foreign Minister, Sergei Lavrov, again warned about the West inciting colour revolutions there. In case of mass demonstrations against the less democratic governments in post-Soviet countries during the coming years, the United States and the EU will again be tempted to show some support. However, their decision-making on how to react to such protests will now have to factor in the risk of triggering a Russian military reaction.

The latter statement may be surprising, but article [35] contains reference to Russian official documents supporting the above statement:

> The new National Security Strategy that President Vladimir Putin signed at the end of 2015 goes further than previous Russian official documents in identifying foreign-sponsored regime change as a security threat. This has been reinforced by the recent experience of Ukraine where Russia accused the United States and the EU of inciting and supporting an anti-constitutional coup. The Russian authorities have voiced concern about "colour revolutions" i.e. toppling governments by mass protests for at least a decade, accompanied by talk of preventing them abroad as well as at home. At the same time, there are signs that Russia is looking into military capabilities and tactics to carry this out. In

> other words, the Russian authorities are moving from securitizing the issue of anti-regime protests to militarizing it. . . . Conflating opposition protests with warfare, extremism and terrorism can be used by Russia to justify military intervention abroad in what others would see as non-military crises.

It seems that Russian authorities react to color revolutions as to the appearance at the international arena of a new kind of weapon, say as starting to use lasers in antimissile defense. The author of the analytic paper [35] concluded:

> This analysis is not to say that Russia's official counter-colour revolution stance is a promise to intervene militarily in all or any protest crisis in its neighborhood. But its official documents and statements indicate how its leaders think about this issue, and this will shape how they react to future events. Any decision would be based on the right combination of Russian interests being in play, and the calculation would vary with each country. . . .
>
> Given the aftermath of the Arab revolutions and the conflict in Ukraine, Western government look like they have lost what enthusiasm they had for people-power revolutions. Nevertheless, as they were with Ukraine's Euro-Maidan, they could still be tempted to react reflexively or Opportunistically—in support of what they would see as pro-democracy protests. And protests remain a possibility across the post-Soviet space, as recent events in Azerbaijan and Moldova show. . . . Either way, Western governments must take seriously the fact that Russia has upgraded its official messaging about colour revolutions and developed justifications for a military response—not least when seen alongside its growing willingness to use force abroad.

Chapter 3

Basics of Physical Lasing

We present schematically the basic scheme of the functioning of physical laser and its basic components (Fig. 3.1). For the moment, we do not consider the functioning of the optical cavity (see Chapters 7 and 8).

3.1 Laser: History of Invention

The invention of laser was one of the main technological outputs of quantum physics.

- The theoretical basis for future creation of laser was implicitly present in the works of Einstein on stimulated absorption and emission (1917).
- In 1939, the Soviet physicist Fabrikant predicted the use of stimulated emission to amplify "short" waves.
- In 1953, Townes, with Gordon and Zeiger, produced the first microwave amplifier, a device operating on principles similar to the laser, but amplifying microwave radiation rather than infrared or visible radiation. Townes's maser was incapable of continuous output.

Social Laser
Andrei Khrennikov

ISBN 978-981-4800-83-9 (Hardcover), 978-0-367-81720-6 (eBook)
www.jennystanford.com

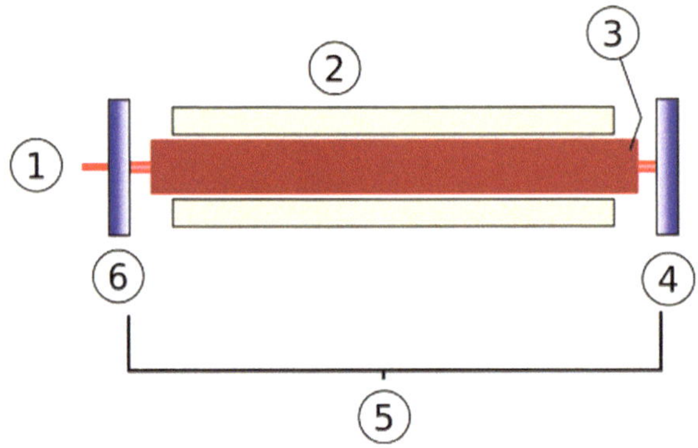

Figure 3.1 Laser's structure: ① output beam; ② pump source (e.g., flashlamp); ③ gain medium; ④ high reflectivity mirrow; ⑥ partially reflective mirrow, ⑤ optical cavity (the block of two mirrows 4 + 6) -laser resonator.

- In 1955, Prokhorov and Basov suggested optical pumping of a multilevel system as a method for obtaining population inversion. Later, the Prokhorov–Basov approach became the main method of laser pumping.

3.2 Spontaneous and Stimulated Emission

We start with a description of the processes of spontaneous and stimulated emission for excited two-level atoms with energy levels $E = E_1, E_2$. Denote the resonance energy of an atom by $E_A = E_2 - E_1$. Typically in the physical literature one proceeds with frequencies and speaks about the resonance frequency $\omega_A = \frac{E_A}{2\pi h}$, where h is the Planck constant. To proceed to the social modeling, we try to eliminate the space-time picture from the model. In particular, we want to operate solely with energies.

We consider the quantum electromagnetic field interacting with atoms and describe the physical processes generated by the interaction.

(1) An atom in the excited state can spontaneously emit a photon. This process is irreducibly random, i.e., even for a single atom

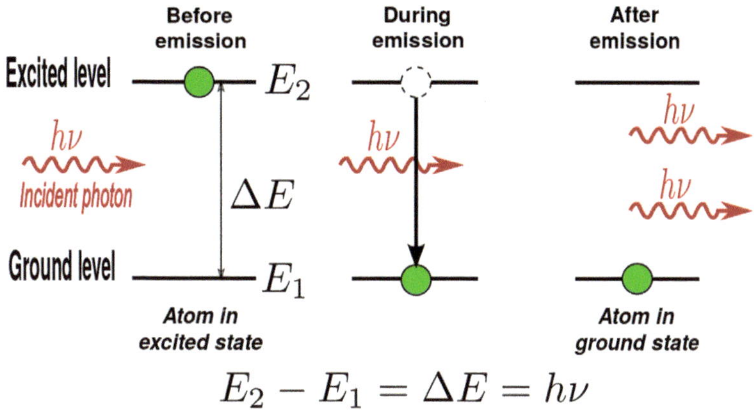

Figure 3.2 Stimulated emission.

it is impossible to predict neither the instance of time nor the direction of the emitted photon.

(2) Atoms in the ground state can absorb from this field only photons having the resonance energy $E_A = E_2 - E_1$.

(3) An excited atom interacting with photons with energy E_A emits a photon of the same energy.

(4) This output flow of photons is *coherent.* All photons produced from the "seed photon" have the same features: direction of flow, polarization, and energy.

Thus one photon produces two. These two interact with two atoms and produce four photons. After n steps this process generates $N = 2^n$ photons. We note that this description and its illustration by Fig. 3.2, although typical for physics textbooks, are too straightforward. In fact, an atom interacts not with a single photon, but with all photons in the field (see section 4.8 and equality (4.4)).

3.3 Population Inversion

The cascade process described above plays a crucial role in generating laser beams carrying huge energy. The main problem is spontaneous emission.[1] Pumping photons into a gain medium (an ensemble of atoms in the ground state) transfers ground-state atoms into excited atoms. But they can spontaneously fall back to the ground level. The basic step in generating lasing is approaching *population inversion*. This is transition from an ensemble of atoms in the ground state to an ensemble in which more than 50% of atoms are in the excited state.

We remark that, for a gain medium composed of two level atoms, population inversion cannot be approached (at least straightforwardly), because the transition probabilities are equal: $p(E_1 \to E_2) = p(E_2 \to E_1)$. These probabilities are known as the *Einstein coefficients.* Their equality for the electromagnetic field can be proven by using thermodynamics for indistinguishable systems following Bose–Einstein statistics. (see section 8.2.1 for the standard physical presentation; see Chapter 6 for a social counterpart of Einstein's theory.) For the electromagnetic field, one needs a gain medium composed of atoms with at least three energy levels, $E_1 < E_2 < E_3$.

The above considerations are about the physical laser based on the quantum electromagnetic field. But, in general, information thermodynamics [168] does not imply the coincidence of the Einstein coefficients for a two-level system. In principle, the social laser can be based on the simplest s-atoms having two levels of social energy.

[1] In fact, the primary problem is that the thermodynamic heat-bath Boltzmann distribution leads to higher population in the lower levels than in the upper levels.

Chapter 4

Basics of Social Lasing

We start considerations with a general foundational discussion. We remark that in quantum theory value a of the observable A obtained in its measurement cannot be interpreted as the property of system S on which the measurement is performed. By the Copenhagen interpretation this value is generated in the complex process of interaction of S and a device used for the measurement of A.

We recall that in the mathematical formalism a (pure) state is represented by the normalized vector ψ of the complex Hilbert space H endowed with the scalar product $\langle\cdot|\cdot\rangle$, i.e., $\langle\psi|\psi\rangle = 1$. So-called mixed states, probabilistic mixtures of pure states, are also considered. They are represented by density operators. The latter state representation plays the basic role in the theory of open quantum systems. A quantum observable is mathematically represented by a Hermitian operator. We shall denote an observable and the corresponding Hermitian operator by the same symbol.

Only if state $\psi \equiv |a\rangle$ of S is an eigenstate of Hermitian operator A representing the observable measured on S, then eigenvalue a can be considered as the property of S (in this eigenstate). For this state, the A-observable takes the value a with probability 1.

Social Laser
Andrei Khrennikov

ISBN 978-981-4800-83-9 (Hardcover), 978-0-367-81720-6 (eBook)
www.jennystanford.com

Here we use the Dirac notation $|a\rangle$ for the eigenstate corresponding to the eigenvalue a, i.e.,

$$A|a\rangle = a|a\rangle.$$

In fact, the label "Copenhagen interpretation" is assigned to a bunch of interpretations of quantum mechanics which can differ essentially [205]. As was proposed by A. Plotnitsky, it is better to speak about interpretations in *the spirit of Copenhagen* (see, e.g., our joint paper [209] for the extended discussion). Generally, quantum theory is characterized by a huge diversity of interpretations. From my viewpoint, this diversity is a sign of the deep crisis in quantum foundations. At the same time, some experts in foundations do not consider the diversity of interpretations as a problem: various interpretations can peacefully coexist. One can choose an interpretation for his own convenience. Other experts would consider this position as a form of scientific opportunism and they badly want to create a single interpretation of quantum mechanics. One of the most successful recent attempts to resolve this foundational crisis was the creation of quantum Bayesianism (QBism), the *subjective probability* interpretation of quantum mechanics (Chapter 11). For physicists, it is difficult (if possible at all) to accept that quantum theory describes the process of decision making about possible outputs of quantum measurements. Moreover, QBism claims that these decisions are made by individual agents, e.g., experimenters. However, QBism provides sufficiently consistent interpretation of quantum mechanics. In particular, in this framework it is possible to proceed without such notions as "quantum nonlocality" and action at a distance.[1]

Once again we emphasize the role of QBism as a bridge between quantum physics and applications of quantum formalism and methodology outside physics, e.g., in sociophysics.

---

[1] I remember the stormy debates after the talk of T. Hänsh at one of the Växjö conferences on quantum foundations. (He is physicist who got the Nobel Prize for experimental laser physics.) When Hänsh was aggressively attacked by representatives of the realist interpretations of quantum mechanics (especially by L. Vaidman) and accused of using QBism, which is totally foreign to physics, he said that he was not able to find another consistent interpretation. He prefers to live with QBism than with a bunch of quantum paradoxes and mysteries.

We presented this overview of the present-day situation in quantum foundations to support our interpretational analysis of the notion of social energy and other emotion-based observables in sociophysics. A few times I was confronted with very sharp opposition for the quantum-like modeling of cognition and social processes. In particular, reviewers can be very critical of operating with the notion of social energy. At the same time, such reviewers are (surprisingly) sure that in quantum mechanics the notion of physical energy can be straightforwardly interpreted. I think that this is not the case. The interpretation of observables, such as energy, spin, polarization, or temperature, for quantum systems is not crystal clear.

Coming back to the interpretations in the spirit of Copenhagen, we point out that this book is based on, so to say, *the genuine Bohr interpretation.* Quantum mechanics is a theory of observations, and the whole experimental arrangement (context) has to be taken into account.

4.1 Social Energy

In physics, the Copenhagen interpretation has to be applied to energy observable E. Although one often says, e.g., "the energy of the electron," the correct meaning of this statement is about the output of the E measurement. The Copenhagen interpretation is well accommodated to the notion of *superposition* of states. Quantum system S can be in state ψ of superposition of two states $|E_1\rangle$, $|E_2\rangle$ corresponding to two different energy levels E_1, E_2

$$\psi = c_1|E_1\rangle + c_2|E_2\rangle, \tag{4.1}$$

where c_i are complex numbers, probability amplitudes, such that $|c_1|^2 + |c_2|^2 = 1$. If system S is in this state, then the probability to get the value E_i of the energy observable is equal to $p_i = |c_i|^2$. Only, for states $|E_1\rangle$, $|E_2\rangle$, the energy can be considered as the property of S.

Supported by the Copenhagen interpretation of the notion of energy in physics, we are ready to consider the very complex notion of *social (mental) energy.* This notion has been actively used in cognition, psychology, social and political science (since the works of James [118–120] and Freud [84, 85], later by Jung [121], see also [166]), and recently in economics and finances, multi-agent modeling, evolution theory, and industrial dynamics ([55, 101, 117], see also [168, 169, 176] for details). And, of course, these previous studies are supporting for our model. However, we emphasize once again that the application of the Copenhagen methodology simplifies and clarifies essentially the issue of the social energy.

4.1.1 Energy of Social Atoms

The simplest quantification of *social energy* (mental energy) can be done by the question (observable) E = "Are you in the state of relaxation or excitement?" This observable takes two values, say $E_1 = 0, E_2 = 1$. The Copenhagen interpretation is strongly involved. Before being asked the E question, s-atom can be in superposition of these two states. Only by confronting the E question the s-atom determines his or her state. The social energy observable can be represented in different forms; for example, in the form of the question E = "Will you go to the demonstration against Trump or Brexit?" Of course, we need not be restricted to the simplest yes–no, to be or not to be, quantification.

Finer quantifications of social energy can be useful as well. Different types of s-atoms can emit and absorb portions of social energy of different magnitudes. Here a type of s-atom is determined by her or his psyche and social environment. In the operational formalism, we can proceed with some grading of the possible social energy levels for s-atoms. Since we restrict consideration to two-level s-atoms, they are characterized by the social energy levels $E = E_1, E_2$, the ground and excited states, and the *resonance energy* $E_A = E_2 - E_1$. We remark that in principle the ground-level energy for one type of s-atoms can be higher than the excitation level energy for another type. Social lasing is possible only for a gain medium with the homogeneous structure of energy levels.

The Copenhagen methodology demystifies the notion of social (mental) energy. (Of course, the same methodology can be applied to any social [mental] observable.) One should not be surprised that the methodology of quantum physics is applicable outside it. The

Copenhagen interpretation presents the very general methodology which is applicable to any kind of measurement. We remark that the use of this measurement methodology does not imply that the whole apparatus of quantum theory can be applicable. One should be careful by checking constraints on the class of systems and observables leading to applicability of one or another part of quantum theory. The quantum-like approach does not mean to copy straightforwardly the complete quantum theory to, say, social science. For example, to derive the *Bose–Einstein statistics*, we have to assume the *indistinguishability of information quanta* and excitations of the quantum information field (see section 4.2).

4.1.2 Energy of the Quantum Information Field

In physics, energy can be assigned not only to atoms, material systems, but also to carriers of interactions, e.g., photons or neutrinos, which are excitations of corresponding quantum fields. (Here "assigned" has the operational meaning: "can be measured.") In social lasing, the interaction processes are formally modeled with the aid of a quantum information field generated by a variety of information sources (see section 4.2). Communications "emitted" by them carry portions of social energy. These quanta of social energy are interpreted as the excitations of the quantum information field.

Again, as in the case of s-atoms, we can proceed with "to be or not to be" quantification of the social energy carried by communications. If an s-atom in the ground state absorbs energy from communication C (and becomes excited), then C carries social energy $E_2 = 1$. If C cannot excite a social atom, then C's energy $E_1 = 0$.

This social energy quantification depends on the concrete ensemble of s-atoms, a social group. It is easy to give examples of social and political communications which would excite the average Englishman or American, but not Russian or Chinese, and vice versa. Thus the definition of an information field's energy is purely operational. In some sense, it is even "more operational" than the definition of the energy for a quantum physical field. It is meaningless to speak about the social energy of the information field without describing the class of "detectors"; in our case, these are s-atoms. As in the case of s-atoms, it is possible to proceed with models based on finer scales of the social energy assigned to excitations of the information field. Each communication C is characterized by social energy E_C. It can be absorbed by an s-atom with the resonance energy $E_A = E_C$.

A variety of communications can carry the same portion of social energy. All communications with $E_C = E$, where E is the fixed portion of the social energy, are considered equivalent from the viewpoint of energy delivering. They can be represented by the same field's state $|E\rangle$. We say that E determines *a mode of the quantum information field: E* is the analogue of the characteristic energy $E_\omega = h\omega$ of the electromagnetic mode with frequency ω.

How many elementary excitations of energy E are carried by the E mode of the quantum information field? This number is determined by the power of information sources emitting communications belonging to the E mode.

4.2 Quantum Field Representation of the Information Flow Generated by Mass Media

The electromagnetic field is the basic energy source for the physical laser. The Bose–Einstein statistics ofexcitations (quanta) of this field plays the crucial role in laser's functioning.

For a social analogue of the physical laser, the basic (social) energy source is the *information field* generated by the mass media and the Internet. As was discussed in section 4.1.2, this field can be considered as composed of information quanta—the field's excitations. These excitations are emitted by a variety of information channels and absorbed by humans, s-atoms. The information field is not a classical physical field defined on the basis of the physical space-time, as, e.g., the classical electromagnetic field given by the vector of the electric and magnetic fields $(E(x), B(x))$, $x \in \mathbf{R}^3$. We model the information field operationally with the aid of creation and annihilation operators, as a quantum field (see section 1.4).

We remark that even in physics only the electromagnetic field has its classical counterpart with the space-time representation. If we consider, for example, the neutrino field, this field has only the quantum representation.

In quantum field theory, a state of a field is mathematically described by a normalized vector belonging to the Fock space—the complex Hilbert space representing superpositions of possible excitations of the field. Consider the E mode of the quantum information field. It represents excitations in the form of communications carrying the portion of social energy E: communications that are graded by the social group under consideration as having social energy E. In the Fock space representation, the quantum state of this field is represented as superposition

$$\Phi = \sum c_n |n, E\rangle = \sum_n \frac{c_n}{\sqrt{n!}} \hat{a}_E^{\star n} |0\rangle, \tag{4.2}$$

where $\sum |c_n|^2 = 1$. Here $\hat{a}_E^{\star}$ is the creation operator for E excitations and $|0\rangle$ denotes the vacuum state of the field—no information excitations. For the real information sources, the sum is always finite. The use of infinite series is the price for mathematical consistency—the possibility of using a Hilbert space.

Generally a social group is exposed to radiation of the quantum information field containing different modes of social energy. It is natural to consider the discrete grading of social energy, $E = E_1, E_2, \ldots, E_n, \ldots$. Here we follow von Neumann, who pointed out [235] that quantum observables with a continuous spectrum are just mathematical idealizations of real measurement procedures. For such discrete grading, the Fock state can be written in the standard form

$$\Phi = \sum \frac{c_k}{\sqrt{k!}} \hat{a}_{E_1}^{\star k_1} \cdots \hat{a}_{E_n}^{\star k_n} \cdots |0\rangle, \tag{4.3}$$

where $k = (k_j)$, and for each multi-index k, only a finite number of k_j differs from zero; the squared sum of absolute values of the coefficients equals to one.

4.3 Coloring Information Excitations

For the quantum electromagnetic field, excitations are photons and a photon's type is determined by index $\gamma = (E, \alpha)$. Here E is photon energy and α encodes additional characteristics linked to lasing coherence of the type of polarization and collimation. In the same way, for the quantum information field, the type of an excitation is determined by index $\gamma = (E, \alpha)$. Here E is *the social energy* carried by an excitation of the information field and α is a *social color* of this excitation encoding its basic social characteristics. The color of an information excitation is linked to the coherence of social actions generated by a social laser. Thus, creation and annihilation operators have to be labeled not by just the energy index E, but by index γ, including the state representation; cf. (4.3):

$$|n, \gamma\rangle = \hat{a}_{\gamma}^{\star n}|0\rangle.$$

In physical lasing γ is the quantum characteristic of a photon. It is based on the observational quantities and indistinguishability of photons (with respect to measurement apparatuses available to experimenters). If it were possible to introduce so-called hidden variables and make photons distinguishable, quantumness would be destroyed (see section 4.6 for a discussion). It would be impossible to approach coherence.

We can mention a few examples of social colors of excitations of the quantum information field: war in Iraq, elections in the USA, Brexit, tsunamis in Japan, sex scandals, anti-globalism, climate change, racism, sexism, Putin, Trump, and so on. Determination of social colors is a sociopsychological phenomenon. It depends on a social group. Thus social colors are not internal characteristics of the information field. They depend on a social group exposed to "information radiation" generated by the mass media.

The social analogue of lasing can be initiated only in a social group, a social gain medium, with a sufficiently *rough coloring structure—to approach high concentration of excitations of the same social color mode in the output beam of excitations.* For example, social color Sex $-$ UK $=$ "the sex scandal of 2017 in the Conservative and Unionist Party (UK)" is appropriate for social lasing. This is the proper social color for the communication C $=$ "A BOMBSHELL dossier naming and shaming 36 Tories suspected of inappropriate sexual behaviour has emerged as Westminster remains engulfed by a sex abuse scandal" ("Express," October 30, 2017.) However, if some s-atom spits this color and starts to operate with 36 colors corresponding to concrete Tory executives involved in the sex scandals, then such an s-atom is not proper for social lasing.

As another example of lasing-friendly social color, we can mention: VA = "crimes of the American Army in Vietnam." Such "fuzzy coloring" played a crucial role in the generation of massive social protests against the Vietnam War in the USA and Europe. However, assigning colors to individual contexts and actors of war crimes[2] destroys color coarse graining and makes social lasing impossible.

SR = "Stalin repressions" is also a lasing-friendly social color. It was widely explored for stimulating a coherent wave of protests against the communist regime in the USSR. However, detalization of the repressions' structure would destroy the strength of this social color. The social energy mode associated with this color, E_{SR}, would be split into a bunch of submodes and coherence of lasing would be destroyed. Detalization of social color SR can include pointing to other people involved in the initiation of the Communist Party cleaning in the 1930s, the internal situation in the country, and the situation in the international arena. If one would start analyzing the causes of Stalin's dictatorship and even the causes of the October Revolution and the First World War, color SR would be weaken even more.

Nowadays, social lasing is often based on symbolic association of the real color to the information flow stimulating emission of social actions. This strategy is widely used in the generation of so-called color revolutions (section 2.2). The modern human gain media are so well designed that it is sufficient just to assign, say, the orange color to information delivered by the mass media and the Internet. Of course, straightforward coloring is not a new strategy in sociopolitical engineering. The Bolshevik propaganda had actively used the red color. In this sense the Great October Revolution was one of the first color revolutions. However, it cannot be treated in the same way as the modern color revolutions. Roughly speaking, it was not a purely color revolution. The red color played an important yet subsidiary role as a complement to the well-developed communist ideology. Nowadays, in color revolutions, color identification often

[2]For example, The My Lai Massacre, Company C, 1st Battalion, 20th Infantry Regiment, 11th Brigade, 23rd (America) Infantry, Lieutenant William Calley Jr., March 16, 1968.

plays a crucial role. As was emphasized in section 2.2, typically color revolutions do not have consistent ideological bases.

4.4 From Rough-Coloring to Indistinguishability

For s-atoms, excitations of the information field carrying the same social energy and color, $\gamma = (E, \alpha)$, are *indistinguishable.* Indistinguishability is the basic condition leading to quantum statistics through thermodynamical analysis for both physical and information systems (see Chapter 5). Indistinguishability of excitations generated by the mass media is relative. It corresponds to coarse graining of coloring and depends on a social group, a collection of s-atoms. If s-atoms do not operate in the indistinguishability regime, i.e., they perform the detailed analysis of the content of communication, then quantum statistical description is inapplicable for them. In this case, social lasing is impossible. We remark that a detailed analysis of the content of communication presumes the use of Boolean logic. Thus, the indistinguishability regime implies deviations from Boolean logic (cf. [163]).

Coarse graining (rough-coloring) generating the indistinguishability regime is the basic feature of human cognition. In particular, since long ago, rough-coloring/indistinguishability has been playing a crucial role in *political propaganda.*

As was stressed in section 1.7, *information overload* is one of the basic reasons for operation in the indistinguishability regime. People simply do not have time and information-processing resources for the deep analysis of the content of communication. They proceed with coarse graining leading to relative indistinguishability. (Here "relative" has the meaning of relative with respect to a social group.) In the modern human society, information overload is combined with complexity of information delivered by a variety of information sources. The majority of population simply does not have the mental capacity to perform a detailed analysis of complex sociopolitical, financial, and ecological problems.

Once again, we remind about the recent wave of color revolutions (section 2.2). Such a revolution is "colored by hands," through the massive assigning of a concrete color to the information flows from the mass media and the Internet, including photos and video streams presenting crowds of people with signs of the concrete color characterizing this "color event" (section 4.3). The content of information is not so much important, only its color. Such information processing is a fruit of today's symbolization and formalization of human thinking, one of the fruits of digitalization.

4.5 The Role of Emotions in Transition to the Indistinguishability Mode: Illustration by Military and Revolutionary Propaganda

The military propaganda during the First World War is one of the first examples of the massive exploring of rough-coloring/indistinguishability at the sociopolitical arena. This propaganda was especially successful in the USA. Here, the task of the pro-war propaganda was the most difficult: it should beat the strong isolationist feeling of Americans. In 1917, in the United States, the Committee on Public Information was established under the leadership of the journalist *George Krill.* In particular, this committee hired Edward Bernays, the Austrian–American pioneer in the field of public relations and propaganda, known as "the father of public relations." Edward Bernays was a relative of Sigmund Freud. He was the first to widely apply Freud's theory of unconscious and psychic energy in social science. Later Bernays referred to this work in the Committee on Public Information as "psychological warfare." Bernays understood well that to approach rough-coloring/indistinguishability, *propaganda has to appeal to emotions* [27, 28]. We also remark that the Bolshevik propaganda was also heavily based on emotions. Some leaders of the Great October Revolutions, such as Lenin and especially Trotsky, were the real masters of the emotion-based propaganda.

In such a regime, instead of a deep analysis of the content of an information excitation carried by communications $\mathcal{C}$, s-atoms operate at an emotional level. As was already emphasized, such information processing implies violation of the laws of classical Boolean logic. To use classical logics, an s-atom should construct a Boolean algebra $\mathcal{B}$ of events. Construction of $\mathcal{B}$ is based on the generation of conjunctions, disjunctions, and negations of all events known by this s-atom. This process is based on a detailed analysis of the contents of events (or communications about these events). It generates distinguishability of components of a rough social color.

In contrast to the American propaganda, during the First World War the German propaganda appealed merely to the logical component. As was commonly recognized, this approach was less successful than the rough-coloring/indistinguishability approach of Bernays. During the Second World War Goebbels's propaganda was emotion based with massive violation of laws of Boolean logics—by using the rough-coloring/indistinguishability approach. In this book, we cannot discuss the interrelation of social lasing and Freud's theory of the unconscious. In Chapter 9 we shall briefly present Freud's model of transformation of different sorts of energy, from physical to psychical (see also monograph [145]).

4.6 Hidden Variables: Genuine Quantum versus Quantum-Like Models

We remark once again that in this book quantumness is understood as an observational (epistemic) feature of systems interacting with measurement devices. The same operational approach is applied to physical, mental, and social processes. The main difference between quantum physics and quantum-like sociophysics is in handling the *hidden variables problem.*

In quantum physics it is claimed that a description of the hidden variables, a description finer than that of the quantum state, is in principle impossible: the quantum state ψ represents complete information about system S in this state. Although this claim is common, its validity was questioned by some authors, such as [145, 154]. However, this critique is typically ignored. There are two main sources for ignoring the counterarguments:

- authority of Bohr,
- quantum technology business.

Niels Bohr, who definitely was the most influential and authoritarian among the founders of quantum mechanics, strongly defended the completeness of this theory. Quantum theory provides the most complete description of the physical phenomena. He wanted to present quantum mechanics as the final theory which cannot be improved. We recall that Einstein did not accept this viewpoint on quantum mechanics. In cooperation with Podolsky and Rosen, he wrote a famous paper [74] questioning Bohr's position. This paper and Bohr's reply [32] initiated a stormy debate. Nowadays the common opinion is that this debate ended in favor of Bohr's position (cf., however, [154]). The main problem of this debate was nonrecognition by debaters of mismatching their philosophic positions:

- Bohr considered quantum mechanics as an epistemic theory, i.e., a theory about knowledge earned from observations.
- Einstein was dreaming for an ontic theory, i.e., a theory about reality as it is.

Einstein dreamed of a purely field theory [75] with the measurement counterpart reproducing quantum mechanics; Bohr was struggling for quantum mechanics as the only possible theory of measurement for microsystems. But both spoke about quantum theory that made the impression that they discuss the same foundational problem. Nowadays this philosophic misunderstanding is complicated by appealing to mathematical no-go theorems.[3] The authors of such theorems (von Neumann, Bell, Kochen and Specker, and others) think that the impossibility of finer description of the quantum phenomena can be proved mathematically. Such no-go mathematics plays a crucial role in modern business with quantum technologies, especially in the justification of the advantages of quantum cryptography and computing.

[3]It seems that neither Bohr nor Einstein was interested in such mathematical statements. Bohr "derived" completeness of quantum mechanics from the existence of the Planck quantum of action, h (the minimal portion of action possible in nature). Einstein completely ignored von Neumann's no-go theorem (by having von Neumann's book on the bookshelf in his Princeton office). Until the last day of his life, he worked on constructing the classical field theory for micro-phenomena, in the spirit of ideas presented in a popular book [75].

By applying the quantum formalism to sociophysics, we have to understand that here Bohr's position is not acceptable. In quantum-like modeling of social processes, we appeal not to the impossibility of a more detailed description, but to psychological factors for ignoring this possibility. However, the ignoring effect is the same as the effect of the impossibility. It is not important whether two systems in the same state are genuinely indistinguishable, i.e., a measurement device which can distinguish them cannot be created in principle, or the possibility to distinguish is not realized, say because an s-atom does not want to analyze the content of a communication or is too busy to do this.

4.7 Coloring Role: Pumping versus Emission

We also point to the following striking similarity in the behavior of atoms and s-atoms. A portion of social energy absorbed by an s-atom generally "lost its color."

Somebody, say Elena, living in Moscow absorbed a social excitation emitted by the Russian radio station *Echo of Moscow*. Typically such excitations have anti-corruption colors. However, this does not mean that her spontaneous relaxation would be directed against corruption. She can emit the portion of social energy absorbed from *Echo of Moscow* in a family scandal or another kind of private or social action.

Such behavior is similar to behavior of atoms interacting with photons. If an atom absorbs a photon carrying some concrete physical characteristics ("color"), such as direction or polarization, then it immediately "forgets" about its pre-absorption value. Later, it can emit a photon with a different color, i.e., a photon flying in the direction different from the direction of the pre-absorption photon, even in the opposite direction.

But we recall that the quantum theory is about observational quantities. We do not know what happens inside the atom between absorption and emission of photons. Nor do we know what happens inside the head between hearing the news from *Echo of Moscow* and going to the kitchen to start a scandal or to the center of Moscow to protest against Putin's policy.

This kind of memory loss can be very useful in approaching population inversion (see [168]). There is no need to pump in a gain medium the social energy of the same color as in laser's output beam. In principle, social energy pumping need not be based solely on information about corruption and other dysfunctions of the government. Shocking news about catastrophes, tornados, and killers is an important part of energy pumping in human gain media.

Now we turn to physics. In contrast to the process of absorption–emission, the process of stimulated emission of photons is characterized by "color" conservation: the "colors" of the emitted photons and photons stimulating emission coincide.

We also point out that in the process of stimulated emission a crucial role is played by the intensity of the stimulating electromagnetic field—the so-called Bosonic effect (see section 4.8, formula (4.4)):

The probability of stimulated emission increases with increase in the intensity of flow bosonic excitations.

In section 4.8, we compare this effect with the *bandwagon effect in psychology.* The operational identity of these two effects, physical and psychological, supports application of the quantum field formalism and methodology to social processes. Stimulated emission of social excitations by excited s-atoms can be considered as exhibition of the bandwagon effect: an s-atom in the excited state exposed by radiation compounded of α-colored social excitations would emit social excitations of the same color.

4.8 Comparing Stimulated Emission in Quantum Physics and the Bandwagon Effect in Psychology and Social Science

We stress that the quantum field description of the stimulated emission is a *collective effect*, i.e., an atom interacts with a bunch of photons and not just with an individual photon. It interacts with all excitations of the quantum electromagnetic field having the resonance energy of this atom, $E_A = E_2 - E_1$.

A crucial role is played by the Bose–Einstein statistics of the photons. We consider the fixed energy (frequency) mode of the quantum electromagnetic field. For fixed color mode α, the n-photon state $|n, \alpha\rangle$ can be represented in the form of the action of the photon creation operator $a_\alpha^\star$ corresponding to this mode on the vacuum state $|0\rangle$:

$$|n, \alpha\rangle = \left[\frac{(a_\alpha^\star)^n}{\sqrt{n!}}\right]|0\rangle \tag{4.4}$$

This representation makes it possible to find the transition probability amplitude from the state $|n, \alpha\rangle$ to the state $|n + 1, \alpha\rangle$ equals to $\sqrt{(n+1)}$. On the other hand, it is well known that the reverse process of absorption characterized by the transition probability amplitude from the state $|n, \alpha\rangle$ to the state $|(n - 1), \alpha\rangle$ equals to $\sqrt{n}$.

Thus, in a quantum Bosonic field, increasing the photon number leads to an increase in the probability of the generation of one more photon in the same state. This constitutes one of the basic quantum advantages of laser-stimulated emission, showing that the emission of a coherent photon is more probable than the absorption.

This behavior of photons, or more generally excitations, of any quantum Bosonic field matches *the cognitive bias known as the bandwagon effect* [56]. This effect is characterized by the probability of individual absorption increasing with respect to the proportion of those who have already done so. People are not interested in the underlying rational justification based on Boolean logic. They "hop on the bandwagon" by taking into account only the number n of those who are already sitting on it. It is important to stress that, for an agent interacting with the bandwagon's population, the personalities of the people on the bandwagon play no role: these people are indistinguishable.

This indistinguishability is only with respect to this concrete interaction: social, financial, racial, gender, or political action characterizing this "bandwagon." Of course, people sitting on the bandwagon are individual agents who can differ crucially: biologically, mentally, culturally.

As we have already emphasized, indistinguishability is a crucial assumption leading to quantum statistics. In the case of the bandwagon effect, this is the Bose–Einstein statistics. Of course, it need not be exactly photon statistics (see section 5.2, formula (5.6)).

Thus, the bandwagon effect can be considered as social exhibition of the Bose–Einstein statistics caused by indistinguishability.

To model social lasing, we consider the information version of the bandwagon effect: an s-atom interacts with the excitations of the information field generated by the mass media and the Internet as well as emitted by other agents. For the concrete social color, these excitations are indistinguishable. And an s-atom in the excited state surrounded by information excitations of social color α emits excitation of the same color with a probability proportional to the number of excitations. A crucial role is played by the field coherence with respect to the concrete color.

We can conclude that the formal mathematical model is the same for physical and social structures. This is the model of stimulated emission of a system, physical or human, interacting with some Bosonic field.

4.9 Social Lasing Schematically

- Each class of information communications is characterized by social energy.
- Each s-atom is characterized by a social energy spectrum.
- Coherence corresponds to social color sharpness, ideally one single α-mode generating a coherent beam of social actions.
- People in the excited state interacting with α-colored excitations of the information field would also emit α-colored excitations.
- The amount of the social energy carried by communications stimulating lasing should match the resonance energy of s-atoms in the human gain medium.
- To approach population inversion, social energy is pumped into the gain medium.
- This energy pumping is generated by the mass media and Internet sources.

- The gain medium should be homogeneous with respect to the social energy spectrum. In the ideal case, all s-atoms in the gain medium should have the same spectrum, $E = E_1, E_2$.[4]
- Social colors of excitations in the energy-pumping beam have no straightforward connection with the social color of excitations in the output beam.
- Information excitations follow Bose–Einstein statistics.
- This statistics matches with the bandwagon effect in psychology.
- The probability of emission of the α-colored excitation by an s-atom in a human gain medium increases with an increase in the intensity of the information field on the α-colored mode.
- This behavior generates a cascade of coherent social actions corresponding to information excitations emitted by s-atoms.

For example, a gain medium consisting of agents in the excited state and stimulated by an anti-corruption coloured information field would "radiate" a wave of anti-corruption protests. The same gain medium stimulated by some information field carrying another social color would generate a wave of actions corresponding to this last color.

[4] In reality, it is impossible to create such a human gain medium. As in physics, the *spectral line broadening* has to be taken into account.

Chapter 5

Information Thermodynamics

Originally, thermodynamics was created to describe the statistical behavior of physical systems. However, its formalism can be applied to arbitrary systems such that it is possible to determine the states corresponding to the fixed values of some quantity, *"energy"* E. This quantity has to be additive with respect to the elements of an ensemble of systems. Both physical and social energies can be considered as the concrete representations of such abstract energy. In this chapter, we present the general scheme of derivation of the basic thermodynamical quantities and relations from statistical mechanics of abstract systems. They can be either physical or social (or arbitrary biological systems, or even AI systems). Classical and quantum statistics correspond to ensembles of distinguishable and indistinguishable systems, respectively (see sections 1.6 and 4.3 for a discussion on indistinguishability of information excitations and its operational meaning).

5.1 Thermodynamics from Combinatorics of State Distribution

In further modeling we shall explore thermodynamics for two types of "information gases":

Social Laser
Andrei Khrennikov

ISBN 978-981-4800-83-9 (Hardcover), 978-0-367-81720-6 (eBook)
www.jennystanford.com

- consisting of information excitations: information transmitters
- consisting of s-atoms: information processors.

The interpretation of thermodynamical considerations in this chapter is used throughout this book. In this book, we are interested in information processing by humans. However, the scheme is very general and it can be applied to a variety of systems, e.g., populations of AI systems or animals.

In fact, E. Schrödinger had already presented the abstract scheme of derivation of the basic thermodynamical quantities from statistical mechanics for ensembles of distinguishable versus indistinguishable systems in his brilliant presentation of statistical thermodynamics [220]. Thus, he did this work for us, although he did not try to apply his approach to nonphysical systems. In short, Schrödinger applied Gibbs' approach based on ideal ensembles of identical systems to quantum systems and obtained quantum statistics, both the Bose–Einstein and Fermi–Dirac cases.

Gibbs's approach to thermodynamics differs crucially from the physical approach of Maxwell and Boltzmann. Gibbs' approach can practically be automatically used in our mental modeling, in contrast to the Maxwell–Boltzmann approach. To proceed with the latter, one has to be sure that the constraints imposed on physical gases can also be assumed for "information gases" (in particular, "social gases"). The validity of these constraints for nonphysical systems is not evident at all. The Maxwell–Boltzmann picture is based on N actually existing physical systems in an actual physical interaction with each other, i.e., gas molecules (or Planck oscillators in the corresponding quantum generalization).

In Gibbs' picture, N identical systems are *mental copies* of the one system under consideration. What could it mean to distribute a given amount of energy between these N mental copies? Schrödinger [220] wrote (p. 3):

> The idea is, in my view, that you can, of course, imagine that you really had N copies of your system . . . Fixing your attention on one of them, you find it in a peculiar kind of "heat bath" which consists of the $N-1$ others.

The crucial point is that both pictures, the physical and mental ones, lead to the same results, the same theory. Moreover, the mental picture is even more easily adapted to the quantum case (the main argument from which Schrödinger's analysis was initiated in [220]). By operating not with an ensemble of real systems but with a mental ensemble, we can apply the methods of statistical thermodynamics not only to physical systems but to systems of any origin, biological, cognitive, social, political, or financial. The details of the output of the Schrödinger–Gibbs derivation of the basic thermodynamical laws can be found in [220]. We present briefly the two basic steps of these derivations endowed with our information terminology.

5.2 Thermodynamics of Distinguishable Systems

There is an ensemble (mental in general) of N identical systems; each of them can be in one of the $\psi_1, \ldots, \psi_k, \ldots$ states characterized by the corresponding energy levels $E_1, \ldots, E_k, \ldots$. The state of an ensemble is determined by the indication that system no. 1 is in a state, say ψ_{j_1}, no. 2 in a state $\psi_{j_2}, \ldots,$ system no. N in a state, say ψ_{j_N}.

Schrödinger [220] was careful in his presentation and he pointed out that one need not assume that a system has the definite value of energy as its objective property. He wrote that this value can be obtained only on "inspection" and measurement. So, we can say that he used the operational approach even in the classical case.

Thus a certain class of states of the ensemble will be indicated by saying that $n_1, \ldots, n_k, \ldots$ of N systems are in the states $\psi_1, \ldots, \psi_k, \ldots$. It is important to remark that $n_1, \ldots, n_k, \ldots$ determine not a single state, but a class of states, precisely because the systems in the ensemble are distinguishable. The number of states belonging to this class is equal to

$$P = P(n_1, \ldots, n_k, \ldots | N, E) = \frac{N!}{n_1! n_2! \ldots n_k! \ldots} \tag{5.1}$$

with the two constraints

$$\sum_i n_i = N, \quad \sum_i n_i E_i = E, \tag{5.2}$$

where N is the number of systems in the ensemble and E is the total energy of this ensemble.

Gibbs' method is based on the following argument. If N is very large (ideally infinitely large), then practically all states will belong to the class of states corresponding to the maximal value of P. By taking into account the constraints (5.2) and using the Lagrange multipliers λ and β for these constraints we obtain

$$\sum_i e^{\lambda-\beta E_i} = N, \sum_i E_i e^{\lambda-\beta E_i} = E. \tag{5.3}$$

We form the *partition function*, the statistical sum over states:

$$Z = \sum_i e^{-\beta E_i}. \tag{5.4}$$

Then the first multiplier can be easily found: $\lambda = \ln(N/Z)$. This is just the normalization factor. The second multiplier plays a more important role. It can be connected with average energy per system in the ensemble:

$$U = E/N = \frac{1}{Z}\sum_i E_i e^{-\beta E_i} = -\frac{\partial \ln Z}{\partial \beta}. \tag{5.5}$$

However, in general the functional dependence $\beta = \beta(U)$ is very complex. Therefore in thermodynamics one assigns direct meaning to β as inverse to temperature, $\beta = 1/T$ (and then typically the average energy is expressed as a function of temperature, $U = U(T)$). The invention of temperature is based on considerations leading to a possibility to operate with a thermometer. In physics, such considerations are justified.

However, in coming applications measurement of information temperature is not well established. We simply do not have good thermometers and procedures for measurement of such a quantity. Proceeding pragmatically we just treat the parameter β as representing the average energy per system in the ensemble of N systems with the energy levels $(E_1, \ldots, E_k, \ldots)$ (see also section 5.3).

By using Eq. 5.3 we also find that the average number of systems per energy level is given by

$$\bar{n}_k = \frac{n_k}{N} = \frac{e^{-\beta E_k}}{Z}. \tag{5.6}$$

It is convenient to represent it in the form

$$\bar{n}_k = -\frac{1}{\beta}\frac{\partial \ln Z}{\partial E_k}. \tag{5.7}$$

This set of equations indicates the distribution of N systems over their energy level Schrödinger [220] remarked (p. 8):

> It may be said to contain, in nutshell, the whole of thermodynamics.

5.3 Thermodynamics of Indistinguishable Systems

Consider a system which is composed of m indistinguishable subsystems. A compound system will be denoted by **S** and its subsystems by S with indexes. It is assumed that an observer can assign to all systems some quantity called energy and satisfy the additivity requirement with respect to its distribution over subsystems of a system. We shall denote the energy of **S** by $\mathcal{E}$ and the energy of S by E. These quantities should not be interpreted as objective properties of systems. Energy $E = E(S)$ can be assigned to S as the output of observation performed by an observer on S. The same interpretation is used for energy $\mathcal{E} = \mathcal{E}(\mathbf{S})$ assigned to **S**.

As always, derivation of thermodynamical quantities from ensemble statistics is started with the construction of *partition function* Z. The possible energy levels of a subsystem S are denoted by $E_1, \ldots, E_j, \ldots$. An energy level $\mathcal{E}_k$ of an m-particle compound system **S** is characterized by a sequence of natural numbers $m_1, \ldots, m_j, \ldots$ of subsystems on the corresponding levels. The latter means "subsystems with measurement outputs" $E_1, \ldots, E_j, \ldots$. Here it is crucial that subsystems with the same value of energy E_j are not distinguished one from another. This indistinguishability determines the form of Z and, hence, all thermodynamical quantities. We have

$$\mathcal{E}_k = \sum_s E_s m_s. \tag{5.8}$$

Thus, the partition function is given by the sum

$$Z = \sum_{(m_s)} e^{-\beta \sum_s E_s m_s}, \tag{5.9}$$

where symbol (m_s) denotes an admissible set of numbers m_s. For the moment, $\beta > 0$ is just a parameter of the model, a Lagrange multiplier.

From $\ln Z$, it is possible to deduce the basic thermodynamical quantities; in particular, the average value of m_s

$$\bar{m}_s = -\frac{1}{\beta}\frac{\partial \ln Z}{\partial E_s}. \tag{5.10}$$

Now, different statistics for systems composed of indistinguishable subsystems can be obtained by considering different possible ranges of values of natural numbers m_s:

(1) $m_s = 0, 1, 2, \ldots$ (Bose–Einstein statistics),
(2) $m_s = 0, \ldots, q$, where $q \geq 1$ is a natural number (parastatistics).

In physics, $q = 1$, i.e., $m_s = 0, 1$; this is the case of the Fermi–Dirac statistics. We remark that in quantum physics selection of the Fermi–Dirac statistics from a bunch of parastatistics is just postulated. Of course, it is confirmed by the experimental situation. One cannot exclude that in social science parastatistics with $q > 1$ can find applications.

By restricting considerations to the Bose–Einstein and Fermi–Dirac statistics and following Schrödinger [220], we find that the corresponding partition functions can be expressed as

$$Z = Z_{\mathrm{BE}} = \prod_s \frac{1}{1 - e^{-\beta E_s}}, \quad Z = Z_{\mathrm{FD}} = \prod_s \left(1 + e^{-\beta E_s}\right). \tag{5.11}$$

This leads to the following basic expression for the average value of m_s:

$$\bar{m}_s = \frac{1}{\frac{1}{\xi} e^{\beta E_s} \mp 1} \tag{5.12}$$

where $0 < \xi \leq 1$ is the so-called parameter of degeneration, $\xi = \xi(m)$. We remark that, for photons, $\xi = 1$, with the Bose–Einstein statistics.

Then one gets the average energy as

$$U = \sum_s \frac{\alpha_s}{\frac{1}{\xi} e^{\beta E_s} \mp 1} \tag{5.13}$$

5.3.1 Social Temperature

In physics, the quantity T inverse to parameter β is interpreted as temperature. In social modeling, we can speak about a kind of *social temperature.* This is a complicated notion and we are not ready to discuss it in detail in this chapter. On one hand, we can try to proceed as in classical thermodynamics. But even by mimicking classical thermodynamics, it is important to remember that such a quantity is of the socioemotional type. Thus it cannot be considered as the objective feature of a social system. One can try to define social temperature through consideration of classes of equivalent thermometers and measurement procedures for social temperature. However, the above thermodynamical considerations for indistinguishable systems represent a quantum situation. Hence, the classical definition of temperature does not match them. As in quantum physics, we can try to introduce a kind of social temperature as a parameter characterizing phase transitions. This is a complicated mathematical theory and we postpone such considerations to one of further publications.

5.3.2 Possible Statistics

The mathematical laws describing the average number of systems are presented in Fig. 5.1. These are very special mathematical laws.

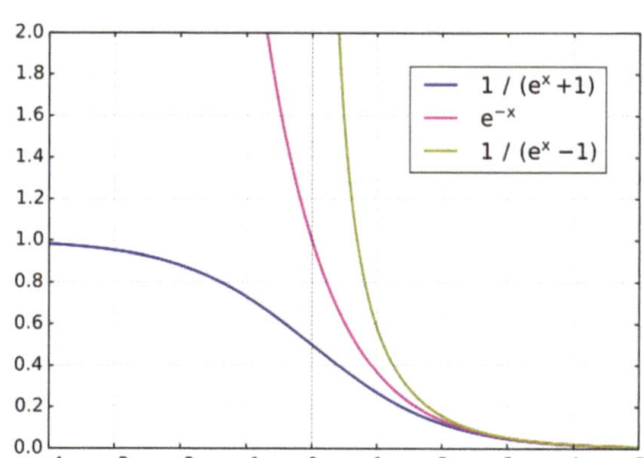

Figure 5.1 Functional dependence for Bose–Einstein, Maxwell–Boltzmann, and Fermi–Dirac statistics.

By considering gases composed of physical systems (Boltzmann–Maxwell approach) one may think that these laws are specified by physics: the Maxwell–Boltzmann law for classical gases and the Bose–Einstein and Fermi–Dirac statistics for quantum gases. However, our derivation (in Gibbs' framework) of formulas (5.6) and (5.12) has nothing to do with physical laws. It is based on consideration of distinguishable versus indistinguishable systems (of any origin). Such a combinatorial derivation makes the origin of the forms of the graphs presented in Fig. 5.1 really mystical. Why is statistical behavior restricted to such forms only? Can one find some statistical behavior with a graph "in between" the graphs corresponding to the Maxwell–Boltzmann, Bose–Einstein, and Fermi–Dirac statistics?[1] What are the additional constraints (different from [in]distinguishability) which can induce new natural forms of statistical dependence on energy? In physics it is proven, both theoretically and experimentally, that only the Maxwell–Boltzmann, Bose–Einstein, and Fermi–Dirac statistics are realizable in nature. However, for social systems, other forms of statistical dependence $f(\epsilon) = \langle n \rangle(\epsilon)$ can also play an important role.

[1]We remark that consideration of the parastatistics does not bring us really new mathematical laws. The parastatistics graphs are simple modifications of the graph for the Fermi–Dirac statistics.

Chapter 6

Thermodynamical Approach to Modeling Population Inversion for Social Laser

We recall that originally Einstein developed theory of spontaneous emission and stimulated emission and absorption by using the thermodynamical approach. At that time the real quantum theory did not exist. The rigorous quantum theory of laser is very complicated. Even in physics, one typically appeals to so-called quasi-classical approach: atoms are quantum and the electromagnetic field is classical (see, e.g., Haken [111]). Since our social lasing theory is oriented to researchers working in social and political science, psychology, and decision making, it is natural to escape appealing to rigorous quantum theory of lasing as long as possible. In this chapter, we plan to mimic Einstein's thermodynamical studies by considering s-atoms and information fields. So, we do not assume the validity of quantum theory. We just emphasize some of its components which would play a crucial role in the thermodynamical derivation of the balance equation for a gain medium composed of s-atoms interacting with an information field.

Social Laser
Andrei Khrennikov

ISBN 978-981-4800-83-9 (Hardcover), 978-0-367-81720-6 (eBook)
www.jennystanford.com

See also section 8.2.1 for the standard physical presentation of Einstein's theory of spontaneous emission and stimulated emission and absorption.

6.1 Einstein Coefficients and Balance Equation for Human Gain Medium Interacting with Information Field

It is postulated that an s-atom imposed to an information field can absorb its excitations and emit them in accordance with Bohr's rule, i.e., the social energy of a field's excitations should match resonance social energy E_{A} of s-atoms. They can also spontaneously emit excitations carrying the amount of social energy that is equal to E_{A}. It is natural to assume that such processes are probabilistic and they are characterized by coefficients having the meaning of *transition probabilities.* The standard mathematical description of such processes is based on master equations. It is important to remark that these are classical master equations, equations for probabilities and not quantum states (density operators). Cf. section 7.3.

In this chapter we proceed with a restricted set of quantum-like assumptions:

(1) discreteness of energy levels of s-atoms
(2) spontaneous and stimulated emission
(3) Bohr's rule
(4) indistinguishability
(5) Bose–Einstein statistics for excitations of information fields

As was shown in Chapter 5, the Bose–Einstein statistics can be derived from indistinguishability.

Now we write the standard master equations for spontaneous and stimulated processes of exchange of social energy. Consider two arbitrary levels of s-atom energy, $E_1 < E_2$, and the resonance energy $E_{\mathrm{A}} = E_2 - E_1$. We start with modeling processes stimulated by the information field. Then transitions between s-atom levels as a result of interaction with the field is described as follows. In an ensemble of s-atoms, if the numbers of s-atoms in these states are given by n_1, n_2, the rate at which stimulated emission occurs is given by

$$\frac{\partial n_2}{\partial t}(t) = -\frac{\partial n_1}{\partial t}(t) = -B_{21}\rho(E_{\mathrm{A}})n_2(t), \qquad (6.1)$$

where $B_{21} \geq 0$ is the transition probability coefficient and $\rho(\epsilon)$ is the energy density of the information field. The rate of emission is thus proportional to the number n_2 of s-atoms in the excited state and to the density of incident excitations of the information field.

In physics (for the electromagnetic field), the proportionality constant B_{21} is known as the *Einstein B-coefficient* for that particular transition, and $\rho(\epsilon)$ is the radiation density of the incident field.

At the same time, we should take into account the process of absorption which removes energy from the field while raising the energy levels of s-atoms from the lower state to the upper state. Its rate is given by an essentially identical equation:

$$\frac{\partial n_2}{\partial t}(t) = -\frac{\partial n_1}{\partial t}(t) = B_{12}\,\rho(E_{\mathrm{A}})n_1(t). \qquad (6.2)$$

The rate of absorption is thus proportional to the number of s-atoms in the lower state n_1 and to the density of the information field.

The assumption of proportionality of the right-hand sides of the equations to the numbers of systems in the corresponding states is standard for any master equation. Hence, here the crucial point is proportionality to the field density which is also natural. However, in principle, especially in applications outside physics, this assumption might be violated. Even in physics this assumption has the restricted domain of validity. If a field interacts with itself, then the nonlinear character of dependence has to be taken into account. This may happen for fields of high intensity or with photons of high energy (such as). Thus by analyzing the behavior of atoms (s-atoms) in high-energy (information) fields, one has to be cautious; modifications to Einstein's theory may be needed. But in this book we proceed with the information analogue of Einstein's theory, i.e., with master equations (6.1), (6.2).

Einstein showed that, for an electromagnetic field (photonic gas), the coefficients of transition are equal, i.e.,

$$B_{12} = B_{21}. \qquad (6.3)$$

In the derivation of this equality, a crucial role is played by the thermodynamical formula (5.12) in setting degeneration parameter $\xi = 1$: the Bose–Einstein statistics for photons. In the present section we proceed under condition (6.3), which essentially simplifies master equations. This condition is natural even for some types of s-atoms: the probability to become excited is equal to the probability to relax and emit an excitation, e.g., in the form of a social action. However, generally, there are no reasons to expect the equality of Einstein's B coefficients for a quantum-like informational field. We consider this problem in section 6.2.

Now we turn to spontaneous emission. The dynamics of the number of excited s-atoms is described by the equation

$$\frac{\partial n_2}{\partial t}(t) = -A_{21} n_2(t), \tag{6.4}$$

where $A_{21} \geq 0$ is the rate of spontaneous emission. In physics (for the electromagnetic field) this constant is referred to as the *Einstein A-coefficient.*

By taking into account both spontaneous and stimulated emission and absorption as well as equality (6.3), we obtain *Einstein's balance equations*:

$$\frac{\partial n_1}{\partial t}(t) = [A_{21} + B_{12}\rho(E_{\mathrm{A}})] n_2(t) - B_{12}\,\rho(E_{\mathrm{A}}) n_1(t), \tag{6.5}$$

$$\frac{\partial n_2}{\partial t}(t) = -[A_{21} + B_{12}\rho(E_{\mathrm{A}})] n_2(t) + B_{12}\,\rho(E_{\mathrm{A}}) n_1(t). \tag{6.6}$$

For the state of equilibrium, we have

$$n_2 A_{21} + n_2 B_{12}\rho(E_{\mathrm{A}}) = n_1 B_{12}\rho(E_{\mathrm{A}}) \tag{6.7}$$

or

$$n_2 A_{21} = (n_1 - n_2) B_{12}\rho(E_{\mathrm{A}}). \tag{6.8}$$

Thus $(n_1 - n_2) \geq 0$. Hence population inversion cannot be approached: the number of s-atoms in the excited state cannot exceed the number of s-atoms in the ground state.

Finally, we make the following important remark. In this section we followed the original Einstein's line of reasoning, i.e., the master equations (6.5), (6.6) are valid for any pair of systems, "atoms," and "fields," which satisfy the system of constraints 1–5. In fact, one need not require the applicablity of the modern quantum theory based on Hilbert space formalism. Thus, although following a modern line of thought, we invented information fields as quantum fields and s-atoms as quantum systems with discrete energy spectra, in principle we might proceed in the "old-fashioned way" as Einstein did, by using thermodynamical arguments and a number of natural constraints on energy levels of systems. In modern quantum physics, such an approach is not so much valuable, since the applicability of the quantum Hilbert space formalism is well tested and therefore commonly accepted. However, in information theory, especially its applications to sociology and politics, economics, and finance, one cannot take the applicability of the standard Hilbert space formalism as granted. Here it is more natural to proceed in the original Einstein manner. We shall continue in the same way in the next section (see section 7.3 for the consistent use of quantum formalism).

6.2 Balance Equation for Steady State and Population Inversion

For an information field, there is no prior reason to select the degeneration parameter ξ (see (5.12)) equal to 1. The latter value corresponds to photon statistics. We proceed with a general information boson field (gas of information excitations) satisfying the thermodynamical constraint (5.12) and assumptions 1–5 at the very beginning of section 6.1.

We know that, for the quantum electromagnetic field interacting with atoms, Einstein's B coefficients are equal. Our aim is to find how they are interrelated—in the presented abstract model of interaction of a field of information quanta with gas of s-atoms with discrete levels of social energy. In physics the standard derivation of the equality (6.3) is based on thermodynamical arguments. One uses the balance between stimulated and spontaneous emissions in the state of equilibrium for the energy exchange between a field and the gas of atoms. In particular, Planck's law for black body radiation plays an important role. We mimic this derivation by using not explicitly the black body radiation formula, but the general thermodynamic formula (5.12). Our final output is the formula (6.17). For $\xi = 1$, it coincides with Einstein's equality (6.3).

The density of the field energy for the state ψ_s corresponding to the social energy level E_s can be represented by the following formula:

$$\bar{E}_s = E_s \bar{m}_s = \frac{E_s}{\frac{1}{\xi} e^{\beta E_s} - 1}. \tag{6.9}$$

As physicists often do [220], we now proceed with the continuous energy parameter ϵ. Then,

$$\bar{\epsilon} = \frac{\epsilon}{\frac{1}{\xi} e^{\beta \epsilon} - 1}. \tag{6.10}$$

If we write the total number of single-photon states with energies between ϵ and $\epsilon + d\epsilon$ as $g(\epsilon) d\epsilon$, where $g(\epsilon)$ is the density of states, then the density of the energy of the field can be written as

$$\rho(\epsilon) = \bar{\epsilon} g(\epsilon) = \frac{\epsilon g(\epsilon)}{\frac{1}{\xi} e^{\beta \epsilon} - 1} = \frac{F(\epsilon)}{\frac{1}{\xi} e^{\beta \epsilon} - 1}, \tag{6.11}$$

where $F(\epsilon) = \epsilon g(\epsilon)$.

Now we consider the gas of s-atoms. By using the equality (5.6), the classical condition of the thermodynamical equilibrium for the gases of distinguishable systems (see section 5.2), we obtain

$$\frac{n_i}{N} = \frac{e^{-\beta E_i}}{Z}. \tag{6.12}$$

As usual, $E_A = E_2 - E_1$ denotes the resonance social energy of an s-atom.

The balance equation for the social energy exchange between s-atoms and the information field has the form (see (6.7))

$$A_{21} e^{-\beta E_2} + B_{21} e^{-\beta E_2} \rho(E_A) = B_{12} e^{-\beta E_1} \rho(E_A). \tag{6.13}$$

$$A_{21} e^{-\beta E_2} + B_{21} e^{-\beta E_2} \frac{F(E_A)}{\frac{e^{\beta E_A}}{\xi} - 1} = B_{12} e^{-\beta E_1} \frac{F(E_A)}{\frac{e^{\beta E_A}}{\xi} - 1}, \tag{6.14}$$

or

$$A_{21} \left(\frac{e^{\beta E_A}}{\xi} - 1 \right) + B_{21} F(E_A) = B_{12} e^{-\beta E_A} F(E_A). \tag{6.15}$$

Since this equation holds for any value of the temperature parameter[1] β, we obtain the system of two equations:

$$-A_{21} + B_{21}F(E_{\text{A}}) = 0, \;\; A_{21} = B_{12}\xi F(E_{\text{A}}). \tag{6.16}$$

Thus

$$\frac{B_{21}}{B_{12}} = \xi. \tag{6.17}$$

In quantum physics for the electromagnetic field, i.e., the gas of photons, the parameter $\xi = 1$. Thus the Einstein equality (6.3) holds. This prevents approaching the population inversion for gain media consisting of two-level atoms. However, in general, for an "information gas," there is no reason to assume that Einstein's B coefficients are equal. We have to consider two cases:

$$\frac{B_{21}}{B_{12}} = \xi \geq 1 \tag{6.18}$$

and

$$\frac{B_{21}}{B_{12}} = \xi < 1. \tag{6.19}$$

The first case is similar to quantum electrodynamics. Here the population inversion for a gain medium composed of two-level s-atoms is impossible. The second case differs drastically from quantum electrodynamics. Here the population inversion is possible even for two-level s-atoms. For such an s-atom (e.g., human), the probability to move from the ground state to the excited state is higher than the probability to emit an excitation, e.g., in the form of a social action. For an s-atom satisfying (6.19), the lifetime of the excited state is longer than the lifetime of the ground state. If

$$\frac{B_{21}}{B_{12}} = \xi \ll 1, \tag{6.20}$$

then such a person can easily become excited by some communication, but she can occupy this state for a sufficiently long time. But, for large populations, it is sufficient that B_{21} is just slightly less than B_{12}.

---

[1]In our purely informational approach this parameter simply characterizes the average energy, so we can say as well "for any value of the average energy."

6.3 Information Laser: The Four-Level Model

The two-level model of laser is the simplest possible. In physics its main problem is that it is difficult to approach population inversion and, hence, to perform stimulated emission of high efficiency. However, as was pointed out in the discussion about the equalities (6.18) and (6.19), for the stimulated emission-absorption of information excitations, the situation might be better and the two-level model of social lasers can have interesting applications. In any event, the lasing model with a gain medium consisting of two-level s-atoms seems to be oversimplified even for s-atoms (e.g., humans). A human can have a more complicated structure of energy levels than just "excited-nonexcited."[2] Moreover, by proceeding in a more general framework and considering mental energy (see section 4.1), it is natural to invent different types of such energy. It is clear that mental energy can take forms different from social energy, at least the social energy as has been interpreted throughout this book—as the energy generating social protests. This is similar to using different types of physical energy, including mechanical, thermal, and radiative.

6.3.1 Radiative versus Nonradiative Emission for Physical Atoms

In physics the best solution of the problem of population inversion is based on the use of the four energy level model. For such a model, a new type of emission (different from spontaneous and stimulated emissions) plays a crucial role. This is *nonradiative emission*. Besides photons, excitations of an electromagnetic field, atoms can emit *phonons*, mechanical excitations.

It is important to note that nonradiative transitions occur if the energy difference between the levels is very small. They typically occur on a much faster time scale than radiative transitions. Thus, for such levels, one can practically (but not completely) ignore

[2]Of course, two-level s-atoms are best for social lasing. Therefore, for realization of such a mechanism of amplification of social actions, it is natural to attempt to transfer s-atoms with complicated social energy spectra into s-atoms with the straightforward structure of excitations, "to be or not to be."

spontaneous radiative emission. Physicists are interested in gain media composed of atoms such that electrons move quickly from a high energy level to a *metastable level* via small nonradiative transitions and then make the final move down to the bottom level via a radiative transition, by emitting a photon.

For further application to information (and, in particular, social) modeling, it is valuable to note that nonradiative transitions of large magnitudes do not occur Frequently, because the crystal structure generally cannot support large vibrations without destroying bonds. But the energy of relaxations is typically not sufficient for the bonds' destruction.

This is a good place to make a remark about the interpretation of excitations (quanta). Phonons, excitations of mechanical motion, are "virtual," rather than real. They are formal mathematical entities that are useful in the quantum-mechanical treatment of atoms in space. Thus, one cannot imagine a phonon as a particle flying in space. This is not possible for photons. The quantum formalism is an *operational formalism*, and excitations are formal entities expressing discrete exchanges of energy. This viewpoint offers a support for our quantum-like social modeling.

6.3.2 Mental Analogues of Radiative and Nonradiative Emissions

Consider a gain medium composed of s-atoms having k energy levels. These s-atoms interact with quantized information fields of two types, say Φ field and ϕ field. The s-atoms and information fields satisfy assumptions 1–5 of section 6.1. The fields do not interact with each other, but interact with s-atoms. For both fields, the energy exchange matches Bohr's rule. For social laser's functioning, it is important that excitations of the Φ field have essentially higher energy than excitations of the ϕ field. In physics, excitations of the first type are photons (radiation) and excitations of the second type are phonons. Thus, transitions between the levels with high transition energy is governed by the Φ field and transitions between the levels with low transition energy is governed by the ϕ field. This "governing" is probabilistic, i.e., for levels with high transition energy, the probability of Φ transition is very high and probability of ϕ transition is very low. For levels with low transition energy, the situation is the opposite.

In social quantum-like modeling, the Φ field can be interpreted as representing information excitations leading to highly energetic social actions such as demonstrations and brutal protest actions. The excitations of the ϕ field lead to actions of lower-energy amplitude, such as tourism, outdoors activities, children, and hobbies. To make a closer analogy with physics, we can proceed on the basis of a more general paradigm of mental energy (section 4.1). We can speak about the two types of mental energy, namely social and private energy. We remark that such specifications of mental energy have an external (observational) character. "Inside an s-atom," the mental energy is not split into different forms. Its specification is expressed only through external actions of s-atoms.

Our aim is to model stimulated emission of the cascade type of coherent excitations of the Φ field. The excitations of the ϕ-field quanta play a subsidiary but very important role. Here the presence of the private activity field ϕ plays a crucial role in the preparation of population inversion, but not in the formation of a coherent output beam of social excitations.

6.3.3 Balance Equation for Steady State and Population Inversion

Consider an s tom with four energy levels such that $E_1 < E_2 < E_3 < E_4$ and transition energies $E_{ij} = E_i - E_j$ satisfying the following condition: E_{43}, E_{21} are essentially less than E_{32}.

The symbol W_{14} denotes the probability of transition from the E_1 level to the E_4 level resulting from absorption of an information excitation of energy E_{41} from the Φ-field pumping. The symbol W_{41} denotes the probability of inverse transition: from the E_4-level to the E_1-level resulting in stimulated emission of Φ excitation of the energy E_{41}. In physics, for a quantum electromagnetic field, these probabilities are equal. This is the equality of the Einstein B-coefficients (see (6.3)). Therefore the standard laser master equations are written for the pump transition probability $W_\text{p} = W_{14} = W_{41}$. The validity of this assumption in our informational model is a complex problem. As was shown in (6.17), in general

$W_{14} \neq W_{41}$ and we proceed in this framework, by setting

$$W_p \equiv W_{14} = \kappa W_{41}, \kappa = 1/\xi, \tag{6.21}$$

where ξ is the degeneration parameter (section 5.3, equality (5.12)).

We also consider the probabilities of transitions from upper levels to lower ones, γ_{ij}, $i > j$, corresponding to spontaneous emission of excitations of both fields. We assume that the probability of transition from the E_4 level to the E_3 level, γ_{43}, is very high, and the next transition probability, γ_{32}, is very small. Thus the E_4 level is metastable. Then, for the next pair of levels, $E_2 \rightarrow E_1$ transition, the probability γ_{21} is again very high. We introduce the time scales of the corresponding transitions, $\tau_{ij}, i > j$. Here τ_{43} and τ_{21} are very short and τ_{32} is essentially longer.

In our information model, the notion of a lifetime is well defined. This is the average time for an s-atom for staying in the state with a concrete degree of excitement. For humans, the lifetime for different levels of mental excitement can be, in principle, measured in experimental cognitive psychology and psycho-physics.

The mental energy structure of such an s-atom can be interpreted in the following way. A person, say Alice, can absorb a strong energetic pulse and move from the ground mental energy state to the state of very high excitement, with energy level E_4. For example, Alice saw some shocking TV news C and approached the state of affect.[3] For Alice, it is very difficult (mentally and even physiologically) to be staying in this state. She wants to relax. In the absence of intermediate levels, she would have only one relaxation possibility: to immediately perform (spontaneous relaxation) some highly energetic social action. Thus, it would be difficult or even impossible (depending on the value of ξ in (6.21)) to approach population inversion. The presence of the intermediate level E_3 gives the possibility to relax partially by "emitting a small portion of mental energy" E_{43} in some private action. State $|E_3\rangle$ is still a state of sufficiently high excitement. But Alice is not overexcited; she can stay in $|E_3\rangle$ for a sufficiently long time. This interplay of the lifetimes, $\tau_{41} \ll \tau_{32}$, plays a crucial role in approaching population inversion.

[3]As was discussed in detail in sections 1.6, 4.3, 4.4 on energetic indistinguishability of communications, the C content does not play a crucial role in generation of this state of high excitement.

In principle, such a three-level structure of an s-atom's mental energy is sufficient for social lasing. However, the presence of the E_2 level is very natural from a psycho-physiological viewpoint. For Alice, who decayed from the state of high excitement $|E_3\rangle$, it is difficult to relax completely and approach immediately the ground state $|E_1\rangle$. It is more natural to consider approaching ground state $|E_1\rangle$ in two steps:

(1) transition from $|E_3\rangle$ to $|E_2\rangle$ with emission of a large portion of social energy, but without complete relaxation;
(2) transition from $|E_2\rangle$ to $|E_1\rangle$ through some private action.

In reality, the coefficients γ_{ij}, transition probabilities, combine emissions of excitations of both information fields, i.e., potentiality for social and private actions. Sometimes the first transition, $|E_3\rangle \to |E_2\rangle$, can be performed through a private action of high amplitude, such as divorce. But the probability of such an event is very small. In gain media which can be used for social lasers, the Φ components of probabilities γ_{43} and γ_{21} are very small compared with the ϕ components; for the coefficient γ_{32}, the situation is the opposite: transition is practically reduced to emission of Φ excitations. In principle, we can ignore the "noisy components" of transitions and couple probabilities γ_{43} and γ_{21} with private actions and probability γ_{32} with social actions.

Now we analyze more carefully interrelation of transitions $|E_3\rangle \to |E_2\rangle$ and $|E_2\rangle \to |E_1\rangle$. Field Φ consists of two energy modes:

(1) the E_4 mode corresponding to energy pumping (by mass-media) into s-atoms
(2) the E_{32} mode corresponding to spontaneous emission of social actions by s-atoms.[4]

So, the pumping E_4 mode cannot excite s-atoms, being in the $|E_2\rangle$ state. Therefore, for approaching population inversion, s-atoms should not stay in the $|E_2\rangle$ state for a long time. We recall that this is the state of essential but not complete relaxation. Of course, the

[4]We remark that, for the moment, we model only the process of approaching population inversion. In the process of stimulated lasing, the E_{32} mode corresponds to the output beam generated in social laser's resonator (see section 7.2).

absolute value of τ_{21}, is not crucial, but the value relative to τ_{32} is. The later quantifies condensation of s-atoms in $|the E_3\rangle$ state. Thus, the quantity τ_{21}/τ_{32} is one of the basic parameters of social lasing.

We are interested in conditions leading to the steady state with population inversion. Of course, they are well known and can be found in textbooks. Our aim is to present them in a nonphysicist-friendly form and without direct coupling to the quantum electromagnetic field and the virtual quantum field representing mechanical vibrations of atoms in physical space. We proceed with two abstract quantum fields, Φ and ψ. We remark that the information interpretation assigned to these fields is not important in formal considerations. It is also useful to remark that analysis below is based on purely classical probabilistic considerations. Some components of "quantumness" are presented in assumptions 1–5 (section 6.1). Equations similar to the one presented below can be written for any four populations using the probabilities of transition from one population to another. We have the following basic balance equation (by dividing it by the total number N of atoms we can write the master equation):

$$\frac{dN_4}{dt} = W_{14}N_1 - W_{41}N_4 - (\gamma_{43} + \gamma_{42} + \gamma_{41})N_4. \tag{6.22}$$

Set $\tau_4 = \frac{1}{\gamma_4}$, where $\gamma_4 = \gamma_{43} + \gamma_{42} + \gamma_{41}$ is the probability of the ϕ transition from the exited state. Here its inverse τ_4 is the lifetime for this state. Then,

$$\frac{dN_4}{dt} = W_{\mathrm{p}}(N_1 - \kappa N_4) - N_4/\tau_4. \tag{6.23}$$

The steady state is determined by the condition $\frac{dN_4}{dt} = 0$:

$$N_4 = \frac{W_{\mathrm{p}}\tau_4}{1 + W_{\mathrm{p}}\kappa\tau_4}N_1, \tag{6.24}$$

where $W_{\mathrm{p}}\tau_4$ is the "normalized pumping rate." If τ_4 is small, i.e., the probability γ_4 is high, then the normalized pumping rate is small and the coefficient of proportionality between the number of excited s-atoms N_4 and the number of s-atoms in the ground state is approximately proportional to this rate:

$$N_4 = W_{\mathrm{p}}\tau_4 N_1. \tag{6.25}$$

Thus, pumping with the coefficient W_p generates the steady-state regime and the number of excited s-atoms is given by (6.24). Pumping with higher probability of excitation generates more excited s-atoms, and a longer lifetime for the excited level has the same effect.

Now we briefly describe the rate dynamics leading to population inversion. We write the balance equations for levels E_3 and E_2:

$$\frac{dN_3}{dt} = \gamma_{43} N_4 - (\gamma_{32} + \gamma_{31}) N_3 \tag{6.26}$$

or

$$\frac{dN_3}{dt} = \frac{N_4}{\tau_{43}} - \frac{N_3}{\tau_3}, \tag{6.27}$$

where $\tau_3 = 1/(\gamma_{32} + \gamma_{31})$ gives the time scale of transitions from the E_3 level to the levels with lower energy. For the steady state generated by pumping, we get

$$N_3 = \frac{\tau_3}{\tau_{43}} N_4. \tag{6.28}$$

Our aim is to have as many s-atoms (information agents, humans) as possible not in the E_4 state of the maximal excitement, but in the metastable E_3 state. Thus, to have a good laser, the condition

$$\tau_3 \gg \tau_{43} \tag{6.29}$$

must be fulfilled. Thus, state $|E_4\rangle$ has to have a very short lifetime, a metastable state. Of course, it is crucial not absolute value τ_{43}, but interrelation given by (6.29). This interrelation is the condition of relative stability of the $|E_3\rangle$ state. Finally,

$$\frac{dN_2}{dt} = \gamma_{42} N_4 + \gamma_{32} N_3 - \gamma_{21} N_2 = \frac{N_4}{\tau_{42}} + \frac{N_3}{\tau_{32}} - \frac{N_2}{\tau_{21}}. \tag{6.30}$$

For the steady state, we obtain

$$N_2 = \beta N_4, \text{ where } \beta = \frac{\tau_{21}}{\tau_{32}} + \frac{\tau_{43}\tau_{21}}{\tau_{42}\tau_3}. \tag{6.31}$$

To approach the population inversion for the pair of levels E_2 and E_3, the coefficient β should be strictly less than 1.

In lasers with good gain medium, the probability of ϕ-field transition from the excited level with the energy E_4 to the level with the energy E_2 approximately equals to zero, so $\tau_{42} \gg 1$. In such a case,

$$\beta \approx \frac{\tau_{21}}{\tau_{32}}. \tag{6.32}$$

Thus, the condition of population inversion has a very simple form:

$$\frac{\tau_{21}}{\tau_{32}} \ll 1. \tag{6.33}$$

We repeat once again that no quantum physics has been involved; these are simple balance equations that are valid for arbitrary populations. The scenarios of the exchange by the fields quanta is determined by the lifetimes of levels.

6.4 Concluding Remark

The model of approaching population inversion presented in this chapter well matches social lasing with pumping information field carrying highly energetic social excitations. For example, excitations generated by shock news: bloody killers, torture (in the ISIS style), wars, etc. A person cannot stay for a long time in the state of excitement generated by such highly energetic information pulses; she partially relaxes to a less energetic state. However, such partial relaxation should not generate an excitation of high social energy.

Chapter 7

Laser Resonator

In physics, the laser resonator in the form of an optical cavity plays a crucial role in the process of amplifying the output beam and making it coherent. We model the functioning of the social laser resonator by "distilling" the physical scheme to exclude the straightforward connection with optics. We proceed with the *quantum master equation for the density operator* describing excitations of the quantum information field inside the resonator of the social laser (see section 7.3). The main aim is establishing proper social interpretations of the basic quantities and parameters of this dynamical system.

The Internet-based *echo chamber* is considered as an important example of social resonators. Its functioning is mathematically represented by the field of social excitations in the form of posts and comments interacting with the human gain medium.

One of the basic consequences of quantum-like dynamics is the existence of the threshold for the pump parameter (section 7.3.5). If this parameter exceeds the threshold, then practically all social energy pumped into the human gain medium is transferred into the output beam of social actions. This is a good place to emphasize once again the role of the huge power of the information flows generated by the modern mass media.

Social Laser
Andrei Khrennikov

ISBN 978-981-4800-83-9 (Hardcover), 978-0-367-81720-6 (eBook)
www.jennystanford.com

In this chapter, we again proceed with gain media composed of two-level s-atoms. Generalization in the case of multilevel s-atoms is straightforward (cf. section 6.3).

7.1 Resonators of Physical Lasers

In laser physics, one of the main problems in creating laser is approaching population inversion. However, population inversion is not enough to generate a laser effect. Stimulated and spontaneous emissions are compete with each other. Thus, before becoming an amplifying device, a gain medium pumped by an external energy source is first radiated as a usual electric "lamp." Here, spontaneous emission is dominating. The light power is distributed over a variety of frequencies and directions of propagation, generally uniformly distributed. It is the optical cavity, the laser resonator, that creates the conditions necessary for stimulated emission to become predominant over spontaneous emission. The cavity or resonator is composed of two mirrors (Fig. 3.1) that *bounce the beam back and forth through the gain medium.* One of the mirrors is only partially reflecting (the left-hand side mirror) and another is totally reflecting (the right-hand side mirror).

In further considerations it is assumed that the gain medium has approached population inversion.

7.1.1 Spontaneous Initiation of Physical Lasing

When functioning of the laser begins, the gain medium emits spontaneously (see Fig. 7.1) in all directions. However, a part of such spontaneous radiation goes along the axis of the optical cavity. These spontaneously emitted photons can travel backwards and forwards between two mirrors (see Fig. 3.1). Thus, over time, as a result of stimulated emission from the gain medium, the energy of the electromagnetic field along the cavity axis, say x, increases very rapidly. At the same time, radiation emitted in directions deviating from the x axis leaves the cavity after a few iterations due to reflection from mirrors with angles different from $\pi/2$.

Another crucial feature of the electromagnetic field propagating along the resonator axis is its coherence in time. Because the

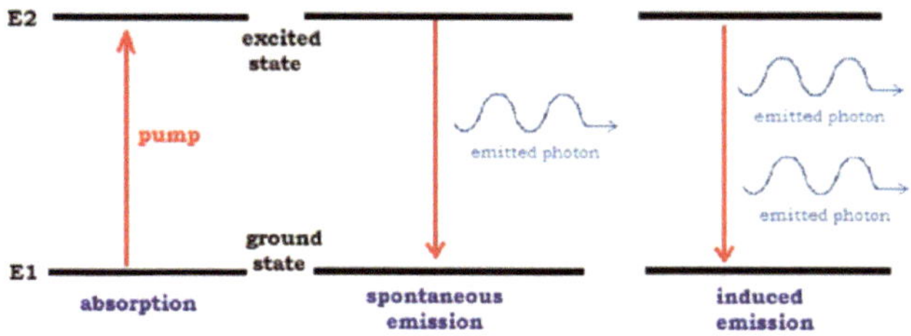

Figure 7.1 Spontaneous emission initiating lasing.

periodic boundary conditions in cavity (with the period $2\pi c/L$, where c is the velocity of light) the electromagnetic x waves interfere constructively. The waves propagating in other directions exhibit destructive interference of some degree.

In the photonic picture we can speak about increasing the number n of photons propagating along the x axis. The rays typically used in the light representation give the directions of the photons' momentum vectors. The concentration of the field inside the cavity increases the probability of stimulated emission rather than spontaneous emission occurring. This is a basic feature of bosons (see section 4.8). We point out that *the behavior of bosons is similar to human behavior, and this is known as the bandwagon effect* (see again section 4.8).[1] In a laser cavity, for each atom, periodically the density of photons surrounding this atom and oriented along the cavity axis becomes very high, resulting in temporal coherence. This makes the probability of stimulated emission of a photon from this atom very high as well. And the emitted photon is directed along the x axis.

7.1.2 Stimulated Initiation of Physical Lasing

Instead of spontaneous emission initiating lasing and filtering improper directions through interference, it is possible to initiate lasing by stimulated emission (see Fig. 3.2). An electromagnetic

[1]This effect is very natural. This form of human behavior was created evolutionally as an important survival factor. At the same time, quantum theory cannot explain the origin of the "bandwagon effect for physical particles," e.g., photons. It only provides a formal mathematical description.

pulse oriented along the x axis and having a frequency matching the energy levels of atoms in the gain medium is injected in the optical cavity. The photons in this pulse interact with atoms and generate stimulated emission along the same axis. It is crucial that these photons have the same phase. One can imagine them as a cloud of increasing size moving between mirrors. Synchronism in movement produces the bosonic stimulated emission effect described above, since each atom is periodically surrounded by an electromagnetic field of high density which is x-oriented.

Of course, spontaneous emission takes place and it can produce photons with momentum vectors different from the x direction. And each of such spontaneously emitted photons can generate a cascade of photons. However, in contrast to the process described in section 7.1.1, there is no need to eliminate such spontaneously emitted photons. They can stay in the cavity, but cascades stimulated by them are very weak compared with the basic cascade generated by the initially injected pulse of photons with momentum x. It is important that photons generated by this pulse propagate coherently and generate the bosonic effect by increasing of the probability of stimulated emission. Photons spontaneously emitted by different atoms as well as the cascades stimulated by these photons are not coherent. And the probability that at the same instant of time a few atoms would spontaneously emit photons in the same direction, say different from x, is practically zero.

It seems that the initializing pulse need not be powerful. One can even find statements such as, to generate a cascade strongly dominating over cascades induced by possible spontaneous emissions, it is enough to coherently inject just two photons. We also remark that, for some gain media, just one or two cascading iterations are sufficient to approach a very high density of photons in the cavity.

7.2 Resonators of Social Lasers

The same competition between spontaneous and stimulated emission plays a crucial role in social processes. People in the excited state may "radiate" social energy spontaneously, say in debates with relatives and friends about political and social problems. Social colors of excitations in such spontaneous radiation are typically randomly distributed, often uniformly distributed. Such emission of social energy cannot lead to coherent social actions.

7.2.1 Structure and Functioning of the Social Resonator

A social resonator consists of a gain medium composed of s-atoms that have already approached population inversion And, say, an Internet-based communication system, e.g., some social network. We call such a system *echo chamber.* We restrict modeling to the Internet-based echo chambers (see also Barbera et al. [20]). Consider the following model.

Each s-atom in the excited state can emit a quantum of social energy in the form of a post or a comment on some post. We call the posts and comments *excitations of the social resonator.* By posting or commenting, i.e., emitting an excitation, an s-atom falls to the ground state. The resonator's excitations play the role of photons in the optical cavity, the resonator of the physical laser. Moreover, to simplify the model, we assume that the social resonator under consideration accepts only excitations of a concrete color, α. This is a strong constraint that is in visible contradiction to the functioning of typical Internet-based social networks. We shall relax it in later modeling. The social color of an excitation plays the role of the direction of propagation of the output beam of photons emitted by the laser resonator, the x axis of the optical cavity.

Suppose that at a fixed instance of time in the social resonator there are n excitations. Each member of the gain medium interacts with all these excitations—with the information field. The boson behavior of excitations implies that the probability that the concrete agent would fall to the ground state and emit an excitation increases with n (section 4.8). It is crucial that if all excitations of the social resonator have a fixed color α, the color of excitations emitted by this agent is also α. This dynamics leads to an exponential increase in the number n of excitations having the α color inside this social resonator (see section 7.3 for the modeling of temporal dynamics). Excitations of colors different from α also can be spontaneously emitted by the gain medium. But they cannot generate the cascade process, since in the present model they are simply blocked.

7.2.1.1 Output beam from the echo chamber

When n becomes sufficiently large (see section 7.3), it is possible to open the output channel of the echo chamber and generate a stable flow of high-intensity excitations of a fixed color α. In "outer space," this flow is realized in the form of meetings, demonstrations, and brutal protest actions.

7.2.1.2 On a spatial picture of quantum physical processes

We remark that in a physical laser resonator, an optical cavity, the light beam bounces back and forth through the amplifying medium. Its temporal dynamics can be represented in the spatial picture, as waves propagating back and forth and reflecting from mirrors. However, this picture of waves propagating in the physical space is merely an artifact of the classical physical model of light. The latter is very useful for visualization of physical processes and it is widely used in laser physics. However, the real theory of lasers is quantum theory.[2] Here, one operates not with classical waves propagating in physical space-time, but with photons, excitations of the quantum electromagnetic field. Although it might be illustratively convenient to treat individual photons as systems moving in space, this picture does not match with quantum theory. A photon's wave function $\psi(t.x)$, $x \in \mathbf{R}^3$, is not well defined. Consequently, by modeling the laser resonator, we have to use particle (quanta) representation and to operate not with spatial waves, but with excitations carrying energy quanta. Such processes are modeled with the aid of creation and annihilation operators (section 7.3.1). The latter mathematical model matches our social applications well. To proceed to the social laser, we have to eliminate the spatial picture from the laser theory.

[2]We remark that in the laser physics community it is sufficiently common to the claim that the functioning of laser can be modeled by means of the semiclassical model [111]: quantized matter and the classical electromagnetic field. However, the semiclassical model is not useful for our social applications, because its rigid coupling to the physical space representation.

Waves in the physical space are good for visualization, but not waves in the information space.

7.2.2 Stimulated Initiation of Social Lasing

As was stressed, the straightforward blocking of excitations with colors different from one fixed color α is a strong assumption. Of course, moderators of social networks block some posts and comments, e.g., having extremist, racist, or sexist content. However, the proportion of blocked excitations seems not to be so high; in any event it is far from 100%. Therefore we have to improve the above model.

As was presented in section 7.1, there are two possible scenarios for initiating lasing:

(1) spontaneous emission and eliminating photons with momentum vectors deviating from the x axis by using the optical cavity
(2) stimulated emission generated by a coherently injected ensemble of photons with the x momentum vector.

In social lasing, the second scenario is preferable, because the social mechanism of elimination of "wrongly directed and spontaneously emitted social excitations" is not so straightforward as in optics (see section 7.2.3).

Thus generation of the beam of social excitations having the same color α is started by injecting a block of α-colored posts into the Internet echo chamber. They are injected at the same moment of time.[3] This initializing block generates a cascade of stimulated emissions. After a few interactions the propagating wave of excitations is so big that the probability of stimulated emission becomes very close to 1. This is a good time to open the output channel of the echo chamber and to transform information excitations into physical social actions.

Of course, spontaneously posted excitations of colors different from α can also be generated in this echo chamber. But they are generated in different moments and have a variety of colors. Even if

[3]By speaking about "the same moment of time," one has to take into account the temporal scale of the echo chamber's functioning. "The same moment" is not the same moment of physical time. It can cover a few days or even weeks.

such a post starts to generate its own cascade, its power is negligible compared with the dominating cascade started with injection of α posts.

7.2.3 Spontaneous Initiation of Social Lasing and Elimination of "Wrongly Colored" Information Excitations

As we have seen in section 7.1, in an optical cavity, direction filtering is done through reflection from mirrors. If the direction of a field's propagation, say u, differs from the x axis of the cavity, then the ray optics implies that soon or later the field radiation will leave the cavity. A bigger difference between directions u and x implies a shorter time needed for escaping from the cavity, or a smaller number of reflections.

We remark again that in the quantum framework, instead of waves propagating in the physical space, we can consider probability waves represented by complex probability amplitudes, normalized vectors of the complex Hilbert space. The inner product determines the angle between two states, two waves of probability.

To be concrete, let us assume that colors of social excitations are labeled by real numbers. Thus α is the real parameter. Now, instead of a fixed color α characterizing the echo chamber, we consider the Gaussian distribution of colors centered at color α with the standard deviation σ. In principle, this distribution of colors in the echo chamber can be very sharp mimicking the fixed color α.

We now consider an information analogue of optical reflection. What does "information excitation $\mathcal{P}$ leaves the echo chamber" mean? It means that the chamber's gain medium does not interact any longer with this post or comment. How can the interaction be terminated? There are two possible ways:

(1) aforementioned straightforward blocking
(2) implicit blocking with the aid of a reflection-like process

So, let $\mathcal{P}$ have color β different from the chamber's color α. If the difference is very large, say $|\beta - \alpha| > 4\sigma$, then the α-colored chamber does not interact with $\mathcal{P}$. In particular, such $\mathcal{P}$ can be blocked by the moderator. Suppose that the difference is not so large, say

$|\beta - \alpha| < 2\sigma$. Then the echo chamber should start to direct this excitation out of it, i.e., the chamber's gain medium should stop interacting with $\mathcal{P}$. This is done through elimination of β color from the echo chamber through posting comments to $\mathcal{P}$ having colors deviating more strongly from α than color β. So, reacting to $\mathcal{P}$ appearance, somebody posts comment $\mathcal{P}_1$ with color β_1, which differs even more from α, $|\beta_1 - \alpha| > |\beta - \alpha|$. Then $\mathcal{P}_1$ can generate posts and comments which differ from α even more. After a few iterations the difference becomes so big, say $|\beta_k - \alpha| > 4\sigma$, that the chamber stops interacting with such posts.

This strategy is actively used in Internet echo chambers. For example, consider the Russian news portals, say, *vz.ru* or *gazeta.ru*. Suppose some news starts to generate antigovernment comments. One can see that such comments are often interrupted by injections of posts on Russian-Ukrainian relations. Suddenly and without any relation to the news, one writes that all "Moscals" are slaves by nature and Ukranians are free people. Immediately in a few comments it is claimed that Ukranians prostitute for America and so on and so forth. The original topic, say the increase in retirement age, is completely forgotten.

7.2.4 Energy Spectrum of the Output Beam: Physical versus Social Lasing

The problem of filtration of the energy of excitations is not less important than the problem of filtration of "wrongly colored" excitations. It is impossible to create a gain medium (either physical or social) with a fixed difference between energy levels.

In physics, the cavity acts as a frequency (energy) filter owing to the numerous round trips: only certain frequencies will be favored, the resonance frequencies of the cavity. In this way, the cavity produces a specific beam of radiation which is composed of a comb of frequencies $\nu_{k_1}, \ldots, \nu_{k_n}$ matching with the basic frequencies of the cavity given by the formula

$$\nu_k = \frac{kc}{2L},$$

where c is the velocity of light and L is the optical length of the cavity, $k = 0, 1, 2, \ldots$. This comb is centered on the spontaneous emission spectrum. The spectral bandwidth of a laser is given by the width of the spontaneous emission.

Typically, a laser is considered the emitter of *monochromatic radiation.* This terminology may make the impression that the output radiation is concentrated just at one fixed frequency, i.e., it emits only photons of fixed energy. However, one has to be careful while operating with the notion of "monochromatic laser." The spectrum of a laser, e.g., helium–neon laser, is monochromatic in the sense that only one color is visible to the naked eye as the line is very narrow. But this is not just one fixed frequency. In reality the physical laser produces a beam with the Gaussian distribution of frequencies (energies) centered on the basic frequency (energy) of the cavity.

In modeling of social processes, we are not interested in the generation of a discrete comb of energies. It is fine to generate a narrow Gaussian distribution of social energies. Echo chambers act as social energy filters: only posts and comments matching the (Gaussian) distribution of energy in the gain medium interact with excited people in this medium and generate new posts and comments. For example, in an anti-globalism echo chamber, posts carrying too high social energy, e.g., to stop the process of globalization of economics by a revolutionary uprising against the capitalist system, would be ignored. They would not generate stimulated emission in the anti-globalism gain medium. Such a gain medium interacts with posts and comments carrying essentially lower portions of social energy. Such a portion is sufficient for stimulating, for example, flight from California to Sweden and participation in an anti-globalism manifestation, but not more. In the same way, a low-energy post proposing to start with yourself and save electric energy at home would be ignored as well. It would not generate new posts and comments.

7.3 Dynamics of the Quantum Information Field in the Social Laser Resonator

We proceed with the standard formalism of the theory of *open quantum systems* by using the *quantum master equation* for the state of a subsystem of a compound system. In our considerations, the latter consists of the quantum information field interacting with the s-atom gain medium. The basic dynamical equation is given by the quantum Markov approximation of the Schrödinger dynamics for the state of the compound system. This is the *Gorini–Kossakowski–Sudarshan–Lindblad* (GKSL) equation [9]. We point out that the using the Markov approximation is an important assumption. Its meaning and validity for social systems is a complex question (see [106] for the corresponding analysis of the general sociopolitical situation). Applicability of the quantum *Markov approximation* to modeling social lasing should be studied in more detail (cf. [184, 185]).

As always, in the theory of open quantum systems, we should extend the notion of the state of a social system by considering mixed states represented by *density operators.*

We follow presentations in physical works (e.g., [82, 83, 102, 212]). One of the essential differences in transition from quantum modeling of real physical processes to quantum-like modeling outside physics is that the Planck constant $\hbar$ cannot be considered as social action quantum. In quantum physics the constant h couples a photon's energy E and angular frequency ω as

$$E = \hbar\omega, \tag{7.1}$$

where $\hbar = h/2\pi$ is the reduced Planck constant. As was already mentioned, we were not able to find a natural social analogue of the angular frequency ω. We tried to proceed without it. The equation (7.1) can be written as

$$\tau = 1/\omega = \hbar/E. \tag{7.2}$$

Here, τ has the physical dimension of time and it can be interpreted as the time scale of the quantum dynamics for the E mode of the quantum electromagnetic field.

While transitioning from genuine quantum to quantum-like modeling, we have to introduce an analogue of the Planck constant, say γ. It is interpreted similarly to (7.2), as a constant coupling the time scale and energy. There is no reason to assume that it is equal to the physical constant $\hbar$. Moreover, we cannot assume that γ is the same for all social processes. If it were the case, it would be really surprising! This is just the time scale of a social process modeled with the aid of the quantum formalism.

7.3.1 Creation–Annihilation Algebras for *s*-Atoms and Quantum Information Field

The quantum information field carrying the fixed amount of social energy E_{F} can be represented in the following operator form:

$$\hat{E}(t) = ue^{-it\frac{E_{\mathrm{F}}}{\gamma}}\hat{a} + \bar{u}e^{it\frac{E_{\mathrm{F}}}{\gamma}}\hat{a}^{\star}, \tag{7.3}$$

where u is the complex field amplitude and $\hat{a}$ and $\hat{a}^{\star}$ are annihilation and creation operators for social excitations of $\hat{E}(t)$, i.e.,

$$\hat{a}|n\rangle = \sqrt{n}|n-1\rangle, \ \hat{a}^{\star}|n\rangle = \sqrt{n+1}|n+1\rangle. \tag{7.4}$$

Here, scaling constants $\sqrt{n}$ and $\sqrt{n+1}$ are selected in such a way that the operator

$$\hat{n} = \hat{a}^{\star}\hat{a} \tag{7.5}$$

can be interpreted as the operator of the excitation number: $\hat{n}|n\rangle = n|n\rangle$.

The creation–annihilation operators satisfy the canonical commutation relation

$$[\hat{a}, \hat{a}^{\star}] = \hat{a}\hat{a}^{\star} - \hat{a}^{\star}\hat{a} = I, \tag{7.6}$$

where I is the unit operator.

The information field Hamiltonian is given by

$$\hat{H}_{\mathrm{F}} = E_{\mathrm{F}}\,\hat{n}. \tag{7.7}$$

The n excitation state $|n\rangle$ is the eigenstate of the field Hamiltonian. In this state the field energy equals nE_{F}. In the absence of interactions, it is preserved in the process of the field's evolution.

As everywhere in this chapter, we consider s-atoms with the two-level structure of the social energy, $E = E_1, E_2$ and transition energy $E_{\mathrm{A}} = E_2 - E_1$. Energy lowering and rising can be formally described by creation and annihilation operators. In contrast to the information field, these are fermionic operators:

$$\hat{b}|E_1\rangle = 0, \ \hat{b}^{\star}|E_1\rangle = |E_2\rangle, \ \hat{b}|E_2\rangle = |E_1\rangle, \ \hat{b}^{\star}|E_2\rangle = 0. \tag{7.8}$$

Hence, they satisfy the fermionic canonical commutation relation

$$\{\hat{b}, \hat{b}^{\star}\} = \hat{b}\hat{b}^{\star} + \hat{b}^{\star}\hat{b} = I. \tag{7.9}$$

These operators are also known as *level lowering and rising operators.* Here the vectors $|E_1\rangle$ and $|E_2\rangle$ represent the ground and excited states, respectively. Similarly to the field's excitations number operator $\hat{n}$, we set $\hat{n}_\mathrm{A} = \hat{b}^\star \hat{b}$. We have $\hat{n}_\mathrm{A}|E_1\rangle = 0$ and $\hat{n}_\mathrm{A}|E_2\rangle = |E_2\rangle$.

We also note that the field and s-atom operators commute:

$$[\hat{a}, \hat{b}] = [\hat{a}, \hat{b}^\star] = [\hat{a}^\star, \hat{b}] = [\hat{a}^\star, \hat{b}^\star] = 0. \tag{7.10}$$

The reader should not be disappointed that to model transitions between s-atom states as well as information field states, we use the same operator algebra as in physics. The formalism of creation and annihilation operators can be used in all models describing transitions between states. For example, besides quantum physics, this formalism plays an important role in analysis of reaction–diffusion equations. This formalism is widely used to model human cognition, decision making, finance, and political studies [14–16, 187].

The Hamiltonian of an s-atom is given by

$$\hat{H}_\mathrm{A} = E_1 + E_\mathrm{A}\hat{n}_\mathrm{A}. \tag{7.11}$$

Since $\hat{H}_\mathrm{A}|E_\alpha\rangle = E_\alpha|E_\alpha\rangle$, $\alpha = 0, 1$, in this model an s-atom that is isolated from the information field and is in the ground state cannot become excited by itself and being in the excited state cannot spontaneously emit a social excitation and relax.

In fact, the forms of the field and s-atom Hamiltonians are selected to preserve the states of concrete social energy (in the absence of interactions). Thus, the forms of these Hamiltonians express the law of energy conservation, but only for the states of concrete energy. The same energy conservation constraint is supposed in quantum physics. Therefore, the reader should not be surprised that we proceed with Hamiltonians of the same form as in physics.

7.3.2 Dynamics of the Compound System *s*-Atom Field

The interaction between the quantum information field and s-atoms is described by Hamiltonian

$$\hat{H}_I = -\gamma g(\hat{b}(t) + \hat{b}^\star(t))\hat{E}(t), \tag{7.12}$$

where the parameter $g > 0$ expresses the strength of coupling between s-atoms and the information field.

The dynamics of s-atom creation and annihilation operators is given by

$$\hat{b}(t) = e^{-it\frac{E_A}{\gamma}}\hat{b}, \quad \hat{b}^\star(t) = e^{it\frac{E_A}{\gamma}}\hat{b}^\star. \tag{7.13}$$

This dynamics is just the Heisenberg picture of the state dynamics given by the Schrödinger equation.

Consider action of the s-atom component of the interaction Hamiltonian at $t = 0$:

$$(\hat{b} + \hat{b}^\star)|E_1\rangle = |E_2\rangle, \quad (\hat{b} + \hat{b}^\star)|E_2\rangle = |E_1\rangle. \tag{7.14}$$

This is the flipping operator representing transitions between states of an s-atom. Such transitions should be compensated by modification of the information field state. Mathematically, this interaction process is represented as composition of two operators.

Formally, we should also use time-dependent creation and annihilation operators in free Hamiltonians $\hat{H}_A$, $\hat{H}_F$. However, these Hamiltonians contain only compositions of operators of the forms $\hat{b}^\star(t)\hat{b}(t)$, $\hat{a}^\star(t)\hat{a}(t)$ and complex exponents containing time dependence cancel each other.

Finally, the interaction Hamiltonian can be represented in the form

$$\hat{H}_I = -\gamma g\Big[ue^{-it\frac{E_F+E_A}{\gamma}}\hat{a}\hat{b} + \bar{u}e^{it\frac{E_F+E_A}{\gamma}}\hat{a}^\star\hat{b}^\star \tag{7.15}$$
$$+ ue^{-it\frac{E_F-E_A}{\gamma}}\hat{a}\hat{b}^\star + \bar{u}e^{it\frac{E_F-E_A}{\gamma}}\hat{a}^\star\hat{b}\Big].$$

The Hamiltonian of the compound system, the quantum information field carrying the social energy E_F and interacting with an s-atom having the social energy spectrum $E = E_1, E_2$, is given by

$$\hat{H} = \hat{H}_A + \hat{H}_F + \hat{H}_I.$$

Denote by ρ the state of this compound system. Its dynamics is described by the von Neumann equation

$$i\gamma\frac{\partial\rho}{\partial t}(t) = [\hat{H}, \rho(t)], \tag{7.16}$$

with the initial condition

$$\rho(t_0) = \rho_A(t_0) \otimes \rho_F(t_0). \tag{7.17}$$

Here, it is supposed that at $t = t_0$ the information states of the s-atom and the field are uncorrelated. Mathematically, this absence of correlations is represented by factorization of the compound system state into the states of the s-atom and the field.

As typical in physical considerations, it is convenient to move to the interaction representation. In this representation, the two components of dynamics, one given by Hamiltonian $\hat{H}_{\text{free}} = \hat{H}_{\text{A}} + \hat{H}_{\text{F}}$ and another by interaction Hamiltonian $\hat{H}_1$, are split. The first part is used for transformation of the density operator $\rho(t)$ of the form $\tilde{\rho}(t) = U^{\star}(t)\rho(t)U(t)$, where $U(t)$ is the unitary parametric group describing the dynamics generated by $\hat{H}_{\text{free}}$, i.e., $U(t) = e^{-i\hat{H}_{\text{free}}/\gamma}$. This is the transformation to the interaction representation. And in the latter state, dynamics is generated by $\hat{H}_1$.

To simplify notations, in the interaction representation we will use the symbol $\rho(t)$ to denote the state (instead of symbol $\tilde{\rho}(t)$). In this representation the von Neumann equation has the form

$$i\gamma\frac{\partial\rho}{\partial t}(t) = \left[\hat{H}_{\text{I}}, \rho(t)\right]. \tag{7.18}$$

This equation can be solved approximately by iterated integration starting with

$$\rho^{(1)}(t) = \rho(t_0) + \frac{1}{i\gamma}\int_{t_0}^{t}\left[\hat{H}_{\text{I}}, \rho(s)\right] ds. \tag{7.19}$$

7.3.3 Gorini–Kossakowski–Sudarshan–Lindblad Equation for the State of the Quantum Information Field

By using the above integral iterations and under some assumptions (see section 7.3.4) we can derive the approximate quantum master equation for the reduced density operator of the field,

$$\rho_{\text{F}}(t) \equiv \text{Tr}_{\text{A}}\rho(\text{t}),$$

where the partial trace is with respect to basis $|E_1\rangle, |E_2\rangle$ in the state space of the s-atom's social energy. This is a special case of the quantum Markov approximation for the dynamics of a subsystem, of a compound quantum system described by the GKSL equation.

Our main task is to present the proper social interpretations for the parameters of the GKSL equation. For s-atoms composing the gain medium, let us denote by T_i, $i = 1, 2$, the average times of being in the ground and excited states, respectively. We can call them the *lifetimes* of relaxation and excitation. Let us denote by r_2 the rate of excitation of s-atoms generated by *social energy pumping* into this gain medium and by r_1 the *absorption rate.*

The quantum master equation has the form

$$\frac{\partial \rho_F}{\partial t}(t) = -A[\hat{a}\hat{a}^\star \rho_F(t) + \hat{a}^\star \rho_F(t)\hat{a} + \text{h.c}] \tag{7.20}$$
$$+ B[\hat{a}\hat{a}^\star \rho_F(t) + 3\hat{a}\hat{a}^\star \rho_F \hat{a}\hat{a}^\star - 4\hat{a}^\star \hat{a}\hat{a}^\star \rho_F \hat{a} + \text{h.c.}]$$
$$- C[\hat{a}^\star \hat{a} \rho_F(t) + \hat{a}\rho_F(t)\hat{a}^\star + \text{h.c.}],$$

where h.c. is the abbreviation for the "Hermitian conjugate." Here,

$$A = r_2(gT_2)^2|u|^2 \tag{7.21}$$

and

$$C = r_1(gT_1)^2|u|^2 \tag{7.22}$$

are *the gain and loss coefficients*, respectively.

These coefficients depend quadratically on the amplitude of the information field u, the interaction coefficient g, and excitation and relaxation lifetimes T_2 and T_1. They are linearly proportional to excitation and absorption rates, r_2, r_1. The term with the saturation coefficient

$$B = r_2(gT_2)^4|u|^4 \tag{7.23}$$

plays a crucial role in generating an exponential increase in the number of excitations in the field. We shall turn to this question later by considering the dynamics of probabilities.

We remark that, since in equation (7.20) the operator coefficients are time-independent, the dynamical state update is not based on the long-term memory. If we consider a discrete time approximation of this dynamics, $\rho_F(t_k)$, $t_k = t_0 + k\delta t$, then the state at the moment t_{k+1} is completely determined by the state at $t = t_k$. Such property is known as the Markov property.

7.3.4 Social Interpretation of Assumptions for Derivation of Quantum Master Equation

There are two basic assumptions for theh derivation of master equation (7.20) from von Neumann equation (7.16) (see [212]). To formulate these two assumptions, it is convenient to introduce the time scale

$$\tau_{\mathrm{F}} = \frac{\gamma}{E_{\mathrm{F}}} \tag{7.24}$$

of the evolution of the field's mode with the energy E_{F} and the time scale

$$\tau_{\mathrm{A}} = \frac{\gamma}{E_{\mathrm{A}}} \tag{7.25}$$

of transition between the states $|E_i\rangle$, $i = 1, 2$.

Assumption 1. Analysis of dynamics (7.20) can be essentially simplified under the following condition:

$$2\pi \frac{\gamma}{E_{\mathrm{A}}} \ll T_2. \tag{7.26}$$

In this situation, the first two terms in the expression (7.15) of the interaction Hamiltonian H_{I} can be neglected in the process of integration with respect to time (see (7.19)).

The inequality (7.26) can be written as

$$\frac{T_2}{\tau_{\mathrm{A}}} \gg 2\pi. \tag{7.27}$$

Thus the lifetime of the excited state of an s-atom should be long enough compared with the transition time scale τ_{A}. We note that the latter is inversely proportional to the resonance energy $E_{\mathrm{A}} = E_2 - E_1$.

For a gain medium with a large gap in social energy, it is easier to satisfy the condition (7.27). For such a social group, the master equation (7.20) gives a better approximation. Of course, the condition (7.27) expresses the complex interplay between the magnitudes of the lifetime for the excited state $|E_2\rangle$ and the size of the energy gap E_{A}.

A good gain medium is characterized by a long lifetime of the excited state and a big gap between the states of relaxation and excitation.

For such a gain medium, the master equation approximation gives the adequate picture of social processes in the gain medium (under additional assumption 2).

Assumption 2. The difference between the energies E_{A} and E_{F} is very small, so the social energy of quanta associated with communications compounding the field does not differs much from the resonance energy of s-atoms; the social energy for transition between s-atom states:

$$|E_{\mathrm{F}} - E_{\mathrm{A}}|\frac{T_2}{\gamma} \ll 1. \tag{7.28}$$

This condition of matching social energies is natural for s-atoms interacting with the information field. The social energy E_{F} carried by excitations of the information field need not be exactly equal to the resonance energy of s-atoms in the gain medium. It can deviate from the latter, but not so much. The right formulation of this statement can be done in probabilistic terms. The emission spectrum of the gain medium has the Gaussian distribution with the mean value E_{F} and sufficiently small standard deviation σ_{A}, where $\sigma_{\mathrm{A}}\frac{T_2}{\gamma} \ll 1$.

Technically, the condition (7.28) justifies the approximation

$$e^{\pm it\frac{E_{\mathrm{F}}-E_{\mathrm{A}}}{\gamma}} \approx 1, \;\; t_0 \leq t \leq T_2, \tag{7.29}$$

in the last two terms of interaction Hamiltonian (7.15). By taking into account that the first two terms in expression (7.15) can be neglected due to **Assumption** 1 interaction Hamiltonian H_{I}, can be approximately treated as a time-independent operator.

The condition (7.28) can be rewritten as

$$|1 - \frac{E_{\mathrm{F}}}{E_{\mathrm{A}}}| \ll \frac{\tau_{\mathrm{A}}}{T_2}. \tag{7.30}$$

Therefore, for the gain medium which is "good" from the viewpoint of **Assumption** 1, i.e., $\frac{\tau_{\mathrm{A}}}{T_2} \ll 1$, the lifetime is long and the social energy gap is high, this condition is satisfied only for very sharp Gaussian distribution, with the mean value E_{F}, of the emission spectrum of the gain medium.

7.3.5 Probabilistic Consequences of the Quantum Markov Dynamics

Now, by using quantum master equation (7.20), we can describe the dynamics of probability $p(t, n)$ to find n excitations in a social laser resonator, e.g., in the form of an Internet echo chamber. We have

$p(t, n) = \langle n|\rho_{\mathrm{F}}(t)|n\rangle$. By averaging Eq. 7.20 with respect to the state $|n\rangle$, we obtain the probabilistic dynamics:

$$\frac{\partial p}{\partial t}(t, n) = -A[(n+1)p(t, n) - np(t, n-1)] \\ - C[np(t, n) - (n+1)p(t, n+1)] \\ + B[(n+1)^2 p(t, n) - n^2 p(t, n-1)]. \quad (7.31)$$

As in physics, we are interested in the steady state of this dynamics, the state of equilibrium in the echo chamber. After approaching this state, the social resonator can be used to emit a powerful wave of social actions. To find a steady state, we set $\frac{\partial p}{\partial t}(t, n) = 0$ and obtain the recurrence equation:

$$p(n) = (A/C)(1 - nB/A)p(n-1). \quad (7.32)$$

In spite of the simplicity of this recurrence equation, its solution cannot be represented in a compact analytic form. In physics, one considers approximations corresponding to different ranges of values of the parameter A/C, *the pump parameter*, representing the interrelation of gain and loss. The equality $A/C = 1$ is *the threshold condition for the laser.*

(1) If $A/C < 1$, then the solution of (7.32) can be approximately represented in the form $p(n) \approx (1 - A/C)(A/C)^n$. Thus, for this region of variation of the parameter A/C, the field is characterized by a small number of excitations: the probability that, in the resonator of the social laser, n excitations can be found decreases exponentially, $p(n) \sim e^{-(\ln C - \ln A)}$.
(2) If $A/C \approx 1$, then the solution has no simple analytical representation. This range of variation of parameter A/C is characterized by large fluctuations of number n of the field's excitations.
(3) If $A/C \gg 1$, then $p(n)$ can be approximated by the Poisson distribution:

$$p(n) \approx e^{-\bar{n}} \frac{\bar{n}^n}{n!}, \quad (7.33)$$

where $\bar{n} = A^2/CB$ is the average number of excitations. It is crucial that the standard deviation of the Poisson distribution $\sigma = \sqrt{\bar{n}}$. This implies that the Gaussian distribution approximating the Poisson distribution with the mean value

$\bar{n} = A^2/CB \gg 1$ is concentrated around $\bar{n}$. Thus (with the high probability) the number of excitations n present in the resonator of the social laser is very large.

In some papers, one sets $A/C = e^b$, i.e.,

$$b = \ln(A/C). \tag{7.34}$$

Then the threshold value equals to zero.

7.4 Concluding Remarks

As in physical lasing, a resonator is the basic component of social lasing. The standard resonator of a physical laser is the optical cavity. Therefore, modeling of the functioning of a physical laser resonator is typically presented in the framework of quantum optics. Our aim was to "distill" the physical scheme from connection with optics. We proceeded with the theory of open quantum systems and the quantum Markov approximation (given by the GKSL equation) of the dynamics of the compound system, s-atoms interacting with the quantum information field. As is typical in quantum-like modeling, we borrowed the mathematical formalism of quantum physics, but assigned new (social) interpretations to the basic quantities and parameters. This interpretational analysis of the (standard) mathematical expressions highlights the following social lasing constraints on the human gain medium and the quantum information field:

- A gain medium is characterized by a long lifetime of the excited state and a big gap between the states of relaxation and excitation.
- The social energy carried by excitations of the information field has to match the resonance energy of s-atoms in the gain medium.
- The interrelation of the magnitudes of the excitation and absorption rates r_2, r_1 and the lifetimes of the corresponding levels T_2, T_1 has to imply the inequality $A > C$, where A and C are gain and loss coefficients, respectively.

- The nonlinear character of interactions between excitations in a social laser resonator (encoded in the B coefficient) plays a crucial role in the initiation of stable social lasing.
- The quantum-like regime of lasing is characterized by the threshold value of the pump parameter b (see (7.34)).

Chapter 8

Correspondence between Notions and Parameters of the Theories of Physical and Social Lasers

In previous chapters we put essential effort to couple the theories of physical and social lasers. Since we hope that at least the first chapters of this book would be readable for non-physicists, we tried to proceed without going deeply into the details of physical laser theory. In this chapter we want to proceed at an advanced physical level. The main aim of this chapter is to ignite the minds of physicists who may be interested in the development of social laser theory. Of course, for experts in laser physics will find the material presented in this chapter to be familiar. However, typically, people working in quantum information and quantum foundations did not study laser theory in detail. Therefore, this chapter can be really useful for them. Our goal is to extract the pertinent features of social protests, including stimulated amplification of social actions, and identify them with the corresponding physical laser concepts (see especially section 8.3). This section contains a brief presentation of the basic ideas covered in the previous chapters of this book.

This chapter is based on the content of the author's work with Zeno Toffano (who is an expert in laser physics) and Francois

Social Laser
Andrei Khrennikov

ISBN 978-981-4800-83-9 (Hardcover), 978-0-367-81720-6 (eBook)
www.jennystanford.com

Dubois [182]. The latter two coauthors independently understood the crucial role of indistinguishability in application of quantum methods at the macroscopic level, in particular for modeling some social phenomena. They developed logical propositional reasoning based on *Eigenlogic*, extending the possibilities of classical logic by giving a fuzzy and multivalued approach using quantum observables [67–69]. Their study on Eigenlogic reasoning matches well with the author's study on social lasing based on quantum-like reasoning based on the indistinguishability of excitations of the quantum information field. The macro-indistinguishability issue was the bridge coupling our research programs [182]. Participation of Zeno Toffano, who is an expert in laser theory, strongly supported research on social lasing. This chapter, especially its technical content, is essentially based on his contribution.

In this chapter, we will build a correspondence of social behaviors with the actual effects observed in physical lasers. The aim is to identify the essential features of laser operation and the associated key physical parameters and connect them (at the formal operational level) with key social parameters in the process of stimulated generation of coherent social actions of social processes. We specify the basic constraints on social gain media and information fields leading to stimulated generation of coherent social actions.

It has to be outlined that at the very beginning of laser research, parallels have been made between laser behavior and other disciplines, considering the laser concept as a general principle with potential applications outside physics, for example in the *synergetics* approach by H. Haken [108–110], where emphasis was laid on laser phase transition and nonlinear dynamics analogy and in the works on laser statistical aspects by M. Lax [192]. On the one hand, the ideas of Haken and Lax support our modeling of information lasers and their applications to social processes. On the other hand, and in contrast with them, we do not only discuss analogies. By establishing the correspondence between the basic features and parameters of quantum physical processes and quantum-like structures, we make a step towards a rigorous theory of social lasing, a part of social engineering.

Following the tradition of physics, we shall operate with frequencies (by keeping in mind their coupling with corresponding energies, physical or social).

8.1 Laser as a Quantum System

On a fundamental level the quantum features at the origin of the laser behavior remain an open question. A full quantum analysis of a laser [111], leads to involved quantum diffusion *Fokker–Planck* equations using quantum operators. But the concept of a laser has been evolving with a better understanding of laser physics. At the beginning, a laser was understood as a source of an intense, sharp beam, but it was soon realized that the high intensity was not the only attribute of a laser; one must also consider its *coherence*, or photon statistics, that constitutes the most fundamental distinction of laser light from the light of a usual *lamp*. The coherence of a laser is more specifically described by first- and second-order correlation functions (see hereafter) which are very popular in quantum optics. One of the important points of practical importance is to determine when a light-emitting device exceeds the so-called threshold and becomes a laser.

8.1.1 Bosonic and Fermionic Creation and Annihilation Operators in Laser Modeling

What are the quantum features that stand at the basis of laser action? This is a complex matter because of the interaction of different quantum systems in a laser. Photons which constitute emitted light interact with energy levels inside the active medium and are also coupled to the cavity.

Photons obey Planck's statistics (Bose–Einsetin statistics for massless particles such as photons) and are represented in quantum field theory by the Bosonic creation operator $a^\dagger$ and annihilation operator a, verifying the commutation relation $[a^\dagger, a] = a^\dagger \cdot a - a \cdot a^\dagger = 1$. The quantum observable for this system is the photon number operator given by

$$n = a^{\dagger} \cdot a, \tag{8.1}$$

which gives for a laser the number of photons in the considered mode.

Radiative transitions in an active medium are between two energy levels: E_1, the lower level, and E_2, the upper level. This constitutes a two-level quantum system and can be treated in quantum mechanics as a fermion-mediated transition obeying the Pauli exclusion principle on each level. Also, here one can introduce creation and annihilation operators respectively for the two levels. So for an energy level E_i one defines the operators $c_i^{\dagger}$ and c_i verifying now an anti-commutation relation: $\{c_i^{\dagger}, c_i\} = c_i^{\dagger} \cdot c_i + c_i \cdot c_i^{\dagger} = 1$.

A full quantum analysis using quantum density operators can be undertaken, and several approaches have been proposed [111]. The result gives a quite complex formulation which is essentially a Fokker–Planck-type equation involving quantum operators. By all these efforts, some simplifying guidelines can be given, quoting Haken [111]:

> The basic quantum mechanical laser equations are strongly reminiscent of the semi-classical laser equations. Indeed, the quantum mechanical laser equations look the same as the semi-classical laser equations with the exception of an additional term which represents fluctuating forces.

8.1.2 Semiclassical Modeling of the Dynamics of the Laser Photon Field

Here, we give the essential models characterizing laser emission. The following equations are obtained by a semiclassical approach for a single-mode laser behavior. One defines two equations for the normalized fluctuations of the photon field b and atomic field α:

$$\delta\dot{b} = -(i\omega - \Delta\omega_{\mathrm{c}})\delta b - i\chi\delta\alpha, \tag{8.2}$$

$$\delta\dot{\alpha} = -(i\varpi - \Delta\omega_{\mathrm{a}})\delta\alpha - i\chi d_0\delta b, \tag{8.3}$$

where

- ω is the cavity resonance angular frequency

- $\overline{\omega}$ is the angular frequency corresponding to the laser transition energy: $\hbar\overline{\omega} = h\nu = E_2 - E_1$; the laser frequency ν is $\overline{\omega}/2\pi$
- $\Delta\omega_c = \tau_p^{-1}$, the cavity bandwidth and τ_p the photon field relaxation time in the cavity
- $\Delta\omega_a = \tau_a^{-1}$, the atomic bandwidth and τ_a is the atomic relaxation time
- d_0 is the unsaturated normalized inversion parameter of the atom population proportional to the difference of population between the two atomic levels (see next section) $d_0 \propto \Delta N = (N_2 - N_1)$
- χ is the field-atom coupling coefficient

Considering the general coherent field amplitudes

$$b = Be^{-i\Omega t}, \ \alpha = Ae^{-i\Omega t}$$

one can solve the two preceding equations and identify the coefficients. The formulation of the field frequency Ω including all contributions is given by an expression comprising all the contributions listed above:

$$\Omega = (\overline{\omega}\Delta\omega_c + \omega\Delta\omega_a)/(\Delta\omega_c + \Delta\omega_a). \tag{8.4}$$

This expression shows that in general the frequency of the laser field does not coincide with the frequency of the cavity. But in general, for a well-designed laser, we have $\Omega \sim \omega \sim \overline{\omega}$; the differences are not significant.

In the field equations given above the amplitude b and its conjugate b^* have their counterpart in the quantum bosonic operators a and $a^\dagger$. The important parameter is of course the photon number $n = |b|^2$, which in a fluctuation analysis will correspond to the average number of photons.

The atomic field α is the counterpart of the quantum fermion operators c and $c^\dagger$. Inversion can be treated quantum-mechanically using the fermion operators corresponding to the two atomic energy levels E_1 and E_2 leading to the corresponding inversion operator [111]:

$$\hat{d} = c_2^\dagger c_2 - c_1^\dagger c_1 \tag{8.5}$$

There is a further simplification which consists in the *adiabatic elimination* of the atomic variations, assuming a selective cavity where $\Delta\omega_c \ll \Delta\omega_a$. This leads to a simplified photon field equation with a Langevin driving fluctuation *noise source*, $F(t) \propto \delta(t)$:

$$\dot{b} = \beta(\Delta - |b|^2)b + F(t), \tag{8.6}$$

where the coefficients are explicitly

$$\beta \sim [4\Delta\omega_c^2/(\Delta\omega_a d_{thr}], \;\; \Delta \sim d_0 - d_{thr}.$$

This last term is the driving parameter which can be considered an order parameter of the laser phase transition. d_0 is the steady- state inversion and corresponds to the mean value $\langle \hat{d} \rangle$ of the quantum atomic inversion operator defined above.

The field equation shows a van der Pol oscillator form, and its interpretation is simple: the nonlinear term in the brackets represents a potential function and is effective only when $\Delta > 0$; that is, when the inversion $d_0 > d_{thr}$ is above the laser threshold. The mean value of the photon field $\overline{b} = \sqrt{\langle b(t)^* b(t') \rangle}$ can be easily derived considering the fluctuation force. This gives

- $\overline{b}_< = 0$ below threshold,
- $\overline{b}_> = \sqrt{n} \neq 0$ above threshold,

where n is the average photon number. This represents, above threshold, a Poisson statistical distribution for the photons with the standard deviation verifying $\Delta n = \sqrt{\langle n^2 \rangle - \langle n \rangle^2} = \sqrt{n}$, showing the important result that the laser effect takes place above threshold with stable intensity.

One can then derive the expression of the laser threshold condition for the inversion parameter given by

$$d_0 > (\Delta\omega_c \Delta\omega_a)/(\chi^2) = d_{thr}.$$

A comment about this last expression is that positive inversion, $d_0 > 0$, is not alone sufficient for laser action, but one must also overcome the losses of the cavity and of the material medium which are globally represented by the term d_{thr}. The region corresponding to $0 < d_0 < d_{thr}$ is the amplifying region where one observes *amplified spontaneous emission.*

8.1.3 Characterization of the Coherence Properties of a Laser Beam with the Aid of Correlation Functions of the First and Second Order

The coherence properties of a laser beam are obtained by averaging the fluctuating quantities. An important difference between *incoherent light* and laser light can be highlighted by higher-order correlation functions. One introduces the first-order correlation function, $g^{(1)}$, and the second-order correlation functions, $g^{(2)}$, as

$$g^{(1)}(\tau) = \left\langle b(t)^{*} b(t+\tau) \right\rangle, \tag{8.7}$$

$$g^{(2)}(\tau) = \; = \left\langle b(t)^{*} b(t+\tau)^{*} b(t) b(t+\tau) \right\rangle. \tag{8.8}$$

For a laser the first-order correlation function well above threshold, $g^{(1)}(\tau)$ can be expressed as a function of the average photon number n and of the phase fluctuations:

$$g^{(1)}_{\text{laser}}(\tau) \sim n \exp(-\Delta\omega_{\phi}\tau), \tag{8.9}$$

$$\left\langle (\phi(\tau) - \phi(0))^{2} \right\rangle \approx 2\Delta\omega_{\phi}\tau, \tag{8.10}$$

where $\Delta\omega_{\phi}$ is the linewidth of the laser field and $\phi(\tau)$ is the *random phase*. The laser behavior above threshold consists in stable intensity, and all the fluctuations are due to the phase. The laser coherence is characterized by its linewidth. One speaks of phase diffusion, and its purest form for a laser corresponds to a Lorentzian line-shape. In this case the phase can be considered a Gaussian random variable.

The second-order correlation function $g^{(2)}$ plays a very important role in the field of quantum optics. It was first studied in the mid-1950s [112]. The quantum theory of light can predict nonclassical values of $g^{(2)}$ that are completely impossible for conventional light waves. The signature of the photon nature of light can have a threefold classification [83] in which the light is described as *bunched, coherent,* or *anti-bunched*. This is summarized for the three cases by

- bunched light: $g^{(2)}(0) \geq g^{(2)}(\tau) \geq 1$
- coherent light: $g^{(2)}(\tau) = g^{(2)}(0) = 1$
- anti-bunched light: $g^{(2)}(0) \leq g^{(2)}(\tau) \leq 1$

Anti-bunched light is only possible in the photon interpretation and is thus another clear signature of the quantum nature of light. It corresponds to sub-Poissonian statistics for photons $\Delta n < n$. Anti-bunched light has practical applications in single-photon sources which are fundamental building blocks in experiments for quantum information. The Hong–Ou–Mandel effect is a two-photon interference effect in quantum optics, which was demonstrated in 1987 [114]: when the photons are identical they will extinguish each other, giving $g^{(2)}(0) = 0$, a clear signature of anti-bunching.

More specifically for a laser, the second-order correlation function $g^{(2)}(\tau)$ increases below threshold from the completely incoherent case, due to spontaneous emission, where $g^{(2)}_{\mathrm{sp}} = 2$, up to a maximum, and then decreases to the limit $g^{(2)}_{\mathrm{las}} = 1$ well above threshold. So for a laser well above threshold the photon statistics variance $(\Delta n)^2$ follows the Poisson behavior equal to its mean value $\overline{n}_{\mathrm{L}}$:

$$(\Delta n)^2 = \overline{n}_{\mathrm{L}} \tag{8.11}$$

The situation below threshold is different and will be discussed more in detail in the next section. In general for less coherent sources than lasers, one has super-Poissonian light with $(\Delta n)^2 > \overline{n}$. The most incoherent situation for photon statistics is thermal blackbody radiation characterized by the Bose–Einstein Planck photon distribution $\overline{n}_{\mathrm{T}}$ at temperature T. In this case it is easily shown [83] that

$$\overline{n}_{\mathrm{T}} = 1/(\exp(h\nu/kT) - 1), \tag{8.12}$$

$$(\Delta n)^2 = \overline{n}_{\mathrm{T}} + \overline{n}_{\mathrm{T}}^2 = \overline{n}_{\mathrm{T}}(1 + \overline{n}_{\mathrm{T}}). \tag{8.13}$$

The first term, $\overline{n}_{\mathrm{T}}$, in the expression of the variance $(\Delta n)^2$ corresponds to the particle nature of light, as in the Poissonian distribution, while the second term, $\overline{n}_{\mathrm{T}}^2$ originates from the thermal fluctuations of the energy of the electromagnetic radiation. This fact was already noticed by Einstein when he introduced the model for the emission and absorption mechanisms involving photons with thermal energy density $\rho_{\mathrm{T}} = (8\pi h\nu^3 c^3)\overline{n}_{\mathrm{T}}$, which has led to the definition of Einstein's coefficients presented hereafter. So, in summary, the first term is a quantum contribution and the second a classical one.

8.1.4 Phase Noise

It is interesting to observe that the problem of the quantum origin of laser light remains an open question even today. One important conclusion from discussions on this problem is that there exists a crucial property of laser light:

A laser must have stable intensity.

More precisely, above threshold intensity, fluctuations become insignificant and only phase noise characterizes the statistics of a laser beam. The simplest quasi-classical model for this behavior is a constant diffusion Fokker–Planck equation for the phase probability distribution $p(\phi)$:

$$dp(\phi)/dt = \Delta\omega/2(d^2 p(\phi)/d\phi^2), \tag{8.14}$$

where $\Delta\omega = 2\pi\Delta\nu$ is the laser linewidth. The difference of behavior below and above threshold has been experimentally verified. A complete analysis based on experiments on semiconductor laser using the Fokker–Planck model has been proposed in [224] showing the phase transition behavior.

In what follows, we will present the expressions for the laser linewidth using the fundamental parameters presented above. The average number of photons n can be related to the output power P using

$$P = 2\Delta\omega_c nh\nu \tag{8.15}$$

for a laser with cavity bandwidth $\Delta\omega_c$ (see discussion on cavities). Then the linewidth below and above threshold can be simply expressed as

$$\Delta\omega_< = 2h\nu\Delta\omega_c^2 P^{-1},\ \Delta\omega_> = h\nu\Delta\omega_c^2 P^{-1}. \tag{8.16}$$

Below threshold the expression of the linewidth scales differently with a factor 2 when compared with the one above threshold. This is due to the additional amplitude fluctuations below threshold. These expressions are called the Schawlow–Townes laws [202] for the laser linewidth showing the laser threshold phase transition behavior. The spectral linewidth for a laser below and above threshold can be put in a very simple form as a function of the photon number n, the cavity characteristics $\Delta\omega_c$, and the second-order correlation function $g^{(2)}(0)$ (as discussed above, this quantity varies from the value 2 to the value 1 when passing the threshold [83, 111, 217]):

$$\Delta\omega = g^{(2)}(0)\Delta\omega_c n^{-1}. \tag{8.17}$$

Similar expressions will be derived in the following section using a system-like approach.

8.2 Laser as a Resonant Amplifier and a Generator: The Role of Positive Feedback

According to Haken in [111] the basic properties of a laser are the following ones:

- high intensity
- high directionality
- spectral purity
- coherence
- laser output as ultra-short pulses

Coherence is a broad concept that encompasses several laser features such as spectral purity but also polarization and directionality. The first question is, what makes laser so special and why is it different from, for example, an ordinary light bulb? What is sometimes confusing is that when considering the acronym LASER, standing for *light amplification of stimulated emission of radiation*, one associates this device with an amplifier, but this is only part of the story. Operating lasers are actually light generators similar to electronic oscillators, where the amplifier is submitted to positive feedback. Feedback is achieved in a laser by using an optical cavity surrounding the active media. Generally, it is a Fabry–Pérot cavity made of up of partially reflecting mirrors as shown in Fig. 3.1.

It is interesting to examine how the transition from noncoherent emission (e.g., ordinary lamp) to that of a laser takes place. Again, according to Haken, [111]

> If we pump the laser only weakly ... the light field consists of entirely uncorrelated individual wave tracks... When we increase the pump power beyond a certain threshold, an entirely new behavior of laser light emerges. It becomes an extremely long wave track.

It has also to be noticed that the laser threshold transition is continuous [224]. Just before threshold, we are in the filtered amplified spontaneous emission regime, sometimes denoted as *superradiant*. From an operational point of view, one has to distinguish between two situations:

(1) A laser as a resonant amplifier: When increasing the pump, the total bandwidth (spectral width) of the light output reduces. One has *amplified spontaneous emission*.
(2) A laser as a generator: When pumping reaches the laser threshold, the gain clamps and all the supplementary pumping energy is transferred to the stimulated emission. The optical power increases linearly with the pump power up to very high levels only limited by the heating of the device. The linewidth decreases as the inverse of the power, giving a highly coherent beam.

8.2.1 Cavity Quality Factor

From a historical point of view the breakthrough of Charles Townes was that of introducing the cavity in the laser. He started, in 1953, with the maser, where he used microwaves corresponding to the ammonia molecule transition [217], and the original idea was to amplify most efficiently microwave signals using cavities.

In order to fully grasp the behavior of a physical laser we will explain here the role of the cavity (see also Chapter 7). A cavity is an optical resonator that is a passive component and so does not by itself amplify energy, but confines and transfers it to selected resonance modes. In order to amplify, one must place the pumped active medium inside the cavity. Also, in this way, a resonant amplifier can become an oscillator for sufficient pumping.

The most common type of lasers uses a Fabry–Pérot cavity. Light bounces back and forth between the mirrors, passing through the gain medium and being amplified each time (see Fig. 3.1). Typically, one of the two mirrors, the output coupler, is partially transparent, and some of the light escapes through this mirror.

Cavities are characterized by their bandwidth $\Delta\omega = 2\pi\Delta\nu$, which is related to the quality factor Q by

$$\Delta\omega/\omega = Q = \omega(\mathcal{E}/P) \tag{8.18}$$

with $\omega = 2\pi\nu$, where ν is the frequency, $\mathcal{E}$ is the energy (unit: joule) stored in the cavity, and P the power lost from the cavity. Because Q can be very high, the cavity bandwidth $\Delta\nu_c$ is typically several orders of magnitude less than the atomic gain bandwidth $\Delta\nu_g$ of the laser material around the two-level transition frequency ν, verifying

$$E_2 - E_1 = h\nu = \hbar\overline{\omega}, \tag{8.19}$$

where E_2 is the upper energy level, E_1 is the lower energy level, and $\overline{\omega}$ is the angular atomic transition frequency. The cavity quality factor Q defined above is one of the important system parameters for the laser. All kinds of losses and also the gain, giving a negative contribution, can be included to get a global quality factor linked to several important laser parameters.

8.2.2 Dynamics of Laser Beam Intensity

Laser operations involve important parameters, first of all the Einstein coefficients for spontaneous emission A, absorption B_{12}, and stimulated emission B_{21} (see also Chapter 6 for social counterpart of Einstein's theory).

The spontaneous emission rate coefficient A is the inverse of the spontaneous lifetime τ_{sp}:

$$A = 1/\tau_{sp} \tag{8.20}$$

The induced effects, stimulated emission and absorption, are characterized respectively by the coefficients B_{21} and $B_{12.}$ Here, in a simplified approach, one does not consider atomic degeneracy. This gives the equality of the induced coefficients $B_{21} = B_{12} = B$. The spontaneous and induced coefficients are related by

$$A/B = 8\pi h\nu^3 c^3, \tag{8.21}$$

where c is the speed of light, h Planck's constant, and ν the frequency.

Rate equations describe the evolution of the atomic populations and photons in the laser process. Considering the intensity of a laser beam $I(\nu)$ as the number of energy quanta $h\nu$ (the photons) produced per second and per unit surface, the rate equation for the photon intensity is given by

$$dI(\nu)/dt = h\nu(N_2 - N_1)Bg(\nu)I(\nu) = c\gamma I(\nu), \tag{8.22}$$

where N_1 and N_2 are the population densities (unit: inverse of a volume) of the energy levels E_1 and E_2 and $g(\nu)$ is the normalized gain profile characterized by the gain bandwidth $\Delta\nu_g$. The solution of the above equation leads, when converting time t to space variation z by the photon velocity c, to an exponential increase or decrease in the laser intensity along the cavity depending on the sign of the gain coefficient γ. Looking at Eq. 8.22, this corresponds to the sign of the inversion term $\Delta N = N_2 - N_1$ in front of the intensity $I(\nu)$.

8.2.3 Laser Oscillation Conditions

Considering a cavity with length L and a coupling mirror reflection factor r, the intensity after one reflection will be multiplied by the loss factor $re^{-\alpha_c L}$. Thus, the condition for sustained power in the cavity is given by the following inequality:

$$\gamma L > \alpha_c L = -Log\, r. \tag{8.23}$$

This condition imposes a minimum gain coefficient, the *threshold gain* $\gamma_{th} = \alpha_c$, which compensates the cavity loss exactly. Because, generally, the reflection factor r is very close to 1, we have $\alpha_c L = -Log\, r \simeq (1 - r)$, the last term being related to the quality factor of the cavity Q_c. Using the definition given in Eq. 8.18, one has the expression

$$Q_c = 2\pi\nu Vol/cS\,(1-r) = 2\pi\nu\, L/c\,(1-r) \simeq 2\pi\nu/c\alpha_c,$$

where $(1 - r)$ corresponds to the fraction of the energy lost by the cavity through one facet, Vol is the cavity volume, and S the cross-section surface. We remark that the cavity length is expressed as $L = Vol/S$.

The cavity bandwidth given by $\Delta\nu_c = \nu/Q_c$ corresponds to the bandwidth of a cavity mode. It decreases when the reflection factor

r approaches 1 (100% of reflection). Another interesting parameter is the photon lifetime in the cavity τ_p, which can be expressed as a function of the cavity quality factor Q_c or also as a function of the cavity bandwidth $\Delta\nu_c$:

$$\tau_p = Q_c/2\pi\nu = 1/2\pi\Delta\nu_c = 1/c\alpha_c. \quad (8.24)$$

Frequencies inside the cavity must verify the constructive interference condition after a round trip in the cavity; they represent the modes of the cavity. The frequency of a mode ν_m, labeled by the number m, and the frequency spacing between modes $\delta\nu$ are respectively given by

$$\nu_m = mc/2L, \delta\nu = \nu_m - \nu_{m\text{-}1} = c/2L. \quad (8.25)$$

One can also estimate the number of allowed modes M in the cavity. This depends on the cavity volume Vol and on the gain bandwidth $\Delta\nu_g$. The mode density ρ_m (per unit volume) is then given by

$$\rho_m = M/Vol = 8\pi\nu^2\Delta\nu_g/c^3. \quad (8.26)$$

8.2.4 Spontaneous Emission, Coherence, and Linewidth

During the amplification process, linewidth is reduced. One can describe this effect by considering the noise introduced by the spontaneous emission coefficient A. The two principal noise sources are the spontaneous emission noise and the thermal noise due to the photons in the cavity—*thermodynamic heath bath*.

Because energy is maintained in the cavity on an average time τ_p, the contribution to the spontaneous energy per mode can be expressed as

$$\epsilon_{sp} = h\nu\ (N_2/\rho_m)\ (\tau_{sp}/\tau_p). \quad (8.27)$$

Thermal noise is characterized by the photon Planck distribution in the cavity, giving the thermal energy contribution per mode of the form

$$\epsilon_T = h\nu/(e^{h\nu/kT} - 1) \simeq h\nu\, e^{-h\nu/kT}.$$

(See Chapter 5 for the general framework including social processes.) Considering that at operating temperatures laser optical frequencies satisfy the inequality

$$h\nu \gg kT$$

(even for a low value of inversion), one can assume that

$$\epsilon_{sp} \gg \epsilon_T.$$

Hence *the spontaneous emission noise is predominant in a laser.*

The total energy stored in the cavity $\mathcal{E}$ verifies the following equation (see also section 6.1) including spontaneous emission:

$$d\mathcal{E}/dt = 2\pi\nu\mathcal{E}/Q_c = Vol\,[h\nu\, B\, g(\nu)(N_2 - N_1)\mathcal{E} + h\nu\, A\, g(\nu)\, N_2]. \quad (8.28)$$

But, using the above expressions, we can transform this equation into the following one:

$$2\pi\nu\,\mathcal{E}\left[Q_c^{-1} - (c\gamma/2\pi\nu)\right] = 2\pi\nu\, Q_c^{-1}\,[N_2/(N_2 - N_1)]\,h\nu.$$

The last term represents the spontaneous emission contribution. One can now estimate the actual linewidth $\Delta\nu$ including the amplification effects:

$$\Delta\nu/\nu = Q_c^{-1} - (c\gamma/2\pi\nu) = Q_c^{-1} - Q_a^{-1}. \quad (8.29)$$

One introduces here the quality factor $Q_a = c\gamma/2\pi\nu$ corresponding to amplification, which is responsible for reducing the linewidth (minus sign). One can also express these quantities as a function of the laser output power P using Eq. 8.18. This leads to the following expression for the linewidth (Schawlow–Townes law) as a function of P, of $\Delta\nu_c$, and of the inversion fraction $n_{sp} = [N_2/(N_2 - N_1)]$:

$$\Delta\nu = 2\pi\ h\nu\, n_{sp}\left(\Delta\nu_c{}^2/P\right).$$

In an ideal laser operating at high power, the inversion fraction tends to $n_{sp} \sim 1$ when $N_2 \gg N_1$. This leads to a very simplified expression for the linewidth:

$$\Delta\omega = 2\pi\,\Delta\nu = P^{-1}\left(h\nu/\tau_p{}^2\right), \quad (8.30)$$

which is equivalent to the one presented earlier in the quantum approach.

8.3 Correspondence between Structures and Parameters of Physical Laser and Information (Social) Laser

In this section we present in a compact form the basic ideas of the previous chapters of this book, but by proceeding at a higher physical level.

Table 8.1 System parameters for physical and information (social) lasers

Laser feature	Phys. laser	Internal par.	Inf. laser	System feature	System par.
pump	inversion	ΔN	exciting inf. field	pump rate	R
amplification	gain medium	g, τ_{sp}	agent gain medium	gain bandwidth	$\Delta \nu_{\text{g}}$
frequency	photon energy	$E_2 - E_1$	social color	mode color	u
lifetime	optical cavity	L, γ	inf. repetition/selection	coherence lifetime	τ_{p}
power	photon number	n	soc. energy rush	power	P
coherence	spectral linewidth	τ_{p}, τ_{sp}	soc. color definiteness	coherence linewidth	$\Delta \nu$

8.3.1 Specification of the Basic Parameters of Physical Laser

A motivation here is to define a laser using system parameters, not relying on specific laser technology. Some laser properties are universal (Haken [108–111]). A tentative correspondence can be made in Table 8.1.

The system parameters constitute a complete system for the rate equations (see section 8.4.2, Eqs. 8.33 and 8.34) and can thus be used for describing laser behavior. The total energy is $\mathcal{E} = n\, h\nu = n(E_2 - E_1)$. The color of the energy is represented by the wavelength $\lambda = c/\nu$, and this is essentially brought in by selection which has its correspondence to the cavity filtering characteristics.

In this way the quantitative character of the information field is represented by the photon number n only. Whereas the social color (section 4.3) of the information is represented by the coherence characteristics, this is actually induced by the filtering/selecting processes due to the cavity (Chapter 7). All coherence characteristics can be put in this category: not only frequency ν but also other properties, such as polarization or beam collimation, which in physical lasers are also achieved by cavity design. One can designate a general parameter for this "coloring" property by α (section 4.3). A quantum state corresponding to an information (social) laser can thus be characterized by an n-photon state in the color mode α, defined by the two quantum numbers n and α. This is actually what we could define as a social information laser mode by assuming single-mode behavior (see also section 4.8):

$$|\Psi_{\text{info}} >= |n, \alpha > . \tag{8.31}$$

In this model, excitations obey Bose–Einstein statistics, and messages with totally different contexts can carry the same portion of social energy (section 4.4). The human gain medium (section 4.9) is characterized by a discrete spectrum of the social energy and the distribution of the social energy in the individual state space follows Fermi–Dirac statistics. An individual can absorb and emit only discrete portions of social energy (adapted to the differences between the energy levels). Individuals belonging to the same gain medium have the same spectrum: stimulated emission of social energy. The basic contribution to the information content is brought in by the laser field and can thus be associated with the photon number n. This quantity can be considered independently from the energy of the atomic transition (whereas the power cannot). Many laser characteristics depend only on n.

8.3.2 General Correspondence between Information and Physical Laser

Here, we want to dress a correspondence between the physical laser and the information laser. The principal characteristics of the information laser are

(1) the quantum information field
(2) its excitations: quanta of information
(3) gain medium represented by human agents

The basic contribution to the information content is brought in by the laser field and can thus be associated with the photon number n. The quantum information field is characterized by several features:

(1) the additivity of social energy
(2) the energy transmitted to an agent by excitations of the information field is absorbed
(3) the indistinguishability of information excitations
(4) the content of the messages is not analyzed, but simply absorbed
(5) the high intensity of information pumping—to approach the threshold

To be able to create a population of individual agents that would function as a kind of laser-emitting wave of social energy, the processes of absorption and emission of social energy have to be discrete. Individual agents have to possess a discrete structure of energy levels. In this model, an agent "falling" from the upper level E_2 to the lower level E_1 cannot stop in between. In modern society, there is no place for those who do not match to the discrete structure of social layers. For absorption, the situation is similar.

For a social equivalent of a laser, the basic source of social energy pumping is the information field generated by mass media and the Internet. In the "particle representation," this field can be considered to be composed of quanta of information. The latter are delivered to us by a variety of information channels such as the mass media, the Internet, social networks, and social apps.

Now we turn to the human gain medium and the (quantum) information field. For simplicity we consider the two-level structure of social energy. Each class of information communications is characterized by the social energy $n(E_2 - E_1)$ carried by n quanta of this field (often in the form of news or news components).

Such communications are also associated with additional characteristics someway linked to the "coherence" of the type of polarization or collimation leading to what can be defined a social color, represented by the general parameter α. For example, α can represent an anti-globalism (or gay-supporting, or antiwar) label. Then this flow generates a "coherent beam" of social actions characterized by the parameter u (e.g., anti-globalism protests). For the human gain medium, such coherence of the output of the stimulated emission process can be easily explained. People in the excited state receiving quanta of information carrying the social color α emit information quanta of the same social color (section 4.8).

The model must also take into account the content of an event. Similarly, in physical reality, an excitation of the quantum electromagnetic field (a photon) is characterized not only by its energy, but also by its internal degree of freedom which is associated with the color parameter α and is principally a cavity effect.

Each class of information communication is characterized by social energy: E carried by n agents. Coherence corresponds to social color sharpness (ideally one single mode u) generating a coherent beam of social actions. People in the excited state receiving quanta of information of the same social color emit information quanta with the same social color. Gain medium consisting of agents in the excited state and stimulated by the anti-corruption colored information field would "radiate" a wave of anti-corruption protests. The same gain medium stimulated by some information field carrying another social color would generate the wave of actions corresponding to this last color.

The correspondence between physical and social laser processes is summarized in Table 8.2.

Table 8.2 Correspondence table of physical and social laser processes

Feature	Physical	Social information
pump	feed power for inversion	massive information flow
cavity	cavity resonator	echo chamber
active medium	2 energy level system	social atoms
field quanta	coherent photons	selected color information

The information field contributing to social energy can be associated with the photon number n, being independent of the frequency. The color of the social energy is, on the other hand, represented by frequency and coherence and is induced by the filtering/selecting process by agents under constant pressure of massive continuous information, leading to resonance effects as in a laser cavity. The massive flow acts as a pump of indistinguishable information accumulating continuously. The result gives stimulated generation of coherent social actions. The atomic transition defines the two-level system energy difference, but the stabilizing character is brought in by the cavity. The color of the social energy is induced by the filtering/selecting process by social agents under constant pressure of massive continuous information, leading to resonance as in a laser cavity.

Here we repeat section 4.8 briefly. We emphasize that the quantum field description of the stimulated emission is a collective effect, i.e., an atom interacts with a bunch of photons and not just with an individual photon. A crucial role is played by the Bose–Einstein statistics of the photons. An n-photon state $|n, u>$, for a fixed parameter α, the color mode, can be represented in the form of the action of the creation operator $a_\alpha^\dagger$ on the vacuum state $|0, \alpha>$:

$$|n, u>= \left[\left(a_\alpha^\dagger\right)^n / \sqrt{n!} \right] |0, \alpha> \tag{8.32}$$

This representation gives the possibility of finding that the transition probability amplitude from the state $|n, \alpha>$ to the state $|n + 1, \alpha>$ equals to $\sqrt{n+1}$. On the other hand, it is well known that the reverse process of absorption characterized by the transition probability amplitude from the state $|n, \alpha>$ to the state $|n - 1, \alpha>$ equals to $\sqrt{n}$. Thus, in a quantum bosonic field, the increase in the number of photons leads to an increase in the probability of generation of one more photon in the same state. This constitutes one of the "quantum advantages" of laser stimulated emission, showing that the emission of a coherent photon is more probable than the absorption.

In social science and finance, the bosonic and fermionic algebras of operators were recently used in works [14–16]. It is interesting that applications outside of physics lead to the use of the operator algebra mixing bosonic and fermionic features in the so-called qubit operator algebra [187] for applications to the theory of decision making.

Finally, we present a citation from the work [201]:

> The concept of a laser has been evolving with both our ever deepening understanding of laser physics and with the progress of various laser technologies. In the very early days of lasers, a laser was understood as a source of an intense, sharp beam. Soon it was realized that the high intensity was not the most fundamental attribute of a laser, rather its coherence or photon statistics that constitutes the most fundamental distinction of laser light from light of a regular lamp. The coherence of a laser is more specifically described by the first- and second-order correlation functions. While such correlation functions are routinely measured in the quantum optics community, most of the opto-electronics (or more generally engineering) community often relies on a few more directly measurable quantities (such as light intensity and/or linewidth) of practical importance to determine when a light emitting device has exceeded the so-called threshold and become a laser.

8.4 Laser Characteristics: Heuristic Pictures

8.4.1 Resonators

Regarding coherence generated by a laser cavity, it is useful to think in terms of "modes," which are the solutions for constructive interference. This is the essential feature brought in by the cavity. A mode relates some particular solution of the degree of freedom of the electromagnetic wave: frequency, polarization (2 degrees of freedom), and spatial configuration of the wave which gives directionality (linked in some way also to frequency because of its dependence on size).

It is always tricky to think of the direction of propagation of photons when we are actually in a "standing wave" configuration in the cavity (one "resolves" the issue by saying that one has two counterpropagating waves). As in most issues of quantum physics, this is a question of "representations," which then turn out to be equivalent. In a laser we face the same "troubles." Actually, no one goes inside the cavity to "observe" in what direction the photons propagate. What is sure is that the "output" possesses these high coherence features, polarization, directionality, and monochromatic light.

Similarly to the processes in the cavity of the physical laser, processes in the social laser resonator are modeled operationally. For example, let us consider the social resonator based on some Internet echo chamber. Although in principle the dynamics of posting can be monitored (even for massive posting), the dynamics of the posts being ignored is unapproachable. Suppose some news collected hundreds of comments. Suppose one can even monitor people opening this Internet page. However, nobody can monitor the reading or ignoring of individual comments.

Now we turn back to the functioning of the physical laser. A "perfect" laser should have all these features at their maximum But, because of physical losses (intrinsic in order to get something out of a laser) and imperfections, we never get optimal conditions. Bad lasers can have some or all of the coherence characteristics degraded (e.g., broad spectrum, large radiation pattern, poor degree of polarization...). The quality can be defined by the coherence time or length measured by the autocorrelation function. For polarization, one can define the degree of polarization by the Stokes parameters on the Stokes sphere (equivalent to the Bloch sphere not using pure density matrices for quantum states).

In the same way it is impossible to create a "perfect" social laser. The principles of social lasing have been known for hundreds (if not thousands) of years. However, those old-fashioned social lasers were not so good as the modern social lasers. In physical lasing, the main problem is degradation of the social coherence characteristics. For example, in the old days, human gain media were characterized by a higher variety of social energy levels (broader social spectrum). Nowadays this variety of social grades is reduced to a few ones. The same can be said about the spectrum of social colors. The speed of reduction of social color diversity can be compared with the speed of reduction of the diversity of species: animals, birds, and insects.

In fact, the situation in modern society is more complicated. Modern social lasing is based on shadowing of people not matching the social energy and color structures supported by the mass media. People with "nonstandard social colors" and social energy patterns are not "visible": the mass media ignores their views. Of course, they can create Internet-based social networks. But networks presenting nonstandard views are typically isolated. One can say that social lasing is impossible without some kind of censure. This was well understood by designers of propaganda during the world wars (see section 4.4).

8.4.2 The Role of the Lasing Threshold

One can be surprised by the analogies with the nonlinear behavior of classical nonlinear oscillators, which is a subject that has been thoroughly studied from the very beginning of the development of electrical circuits. For example, the van der Pol oscillator has similar nonlinear characteristics.

Nevertheless, there are some "heuristic" guidelines which seem to be universal. The nonlinear gain plays a crucial role (also in electronic oscillators). It is a sort of stabilizer that permits the system to go through the transition smoothly and then stabilizes its amplitude. A very simple way to grasp the threshold behavior is from the "depurated" rate equations. A *two rate-equation model* can be used. The parameters in use are the following:

- pump power R
- light output power P (proportional to the photon number n)
- the inversion of the gain medium ΔN, proportional to the population difference $N_2 - N_1$

This simple nonlinear model enables us to show the laser phase transition through the equations

$$d\Delta N/dt = R - \Delta N/\tau_{\mathrm{sp}}, \tag{8.33}$$

$$dP/dt = g\,(P \cdot \Delta N) - P/\tau_{\mathrm{p}}. \tag{8.34}$$

This equation system is nonlinear, because of the product term $(P \cdot \Delta N)$ in the second equation. The global behavior is of the Van der Pol oscillator type, as already discussed earlier in the chapter. These figures show the laser threshold phase transition. Laser modeling may be carried out at different approximation levels. Steady-state solutions of the system of Eqs. 8.33 and 8.34 lead to approximate expressions of laser power function of the pump rate holding below, through, and above threshold [224]. These rate equations can be considered universal for laser operation [77].

We remark that one has to be careful with the use of the formal picture of laser as a physical or social system described by Eqs. 8.33 and 8.34. Appealing to quantum theory and in particular to the quantum master equation (section 7.3) illuminates the internal characteristics of systems leading to the above equations. For example, a crucial role is played by the indistinguishability of the laser field excitations (the electromagnetic field or the information field) as well as the elements of the gain medium (atoms or s-atoms). Indistinguishability leads to quantum statistics playing a crucial role in the generation of stimulated emission of the cascade type. Thus, although for social lasing it is sufficient to create any social system with characteristics satisfying Eqs. 8.33 and 8.34, quantum-like modeling is an important tool for clarifying the internal structure of social laser design.

Chapter 9

Freudian Approach to Psychic Energy

This chapter can be considered as a historical illustration of studies on the notion of psychic (mental energy). We present the views of Sigmund Freud. They are heavily based on the split of psychic processes into unconscious and conscious counterparts. Freud's approach can be considered as an attempt to go beyond the operational definition of psychic (mental) energy (section 4.1). We are especially interested in the following two components Freud's theory:

(1) unconscious representation of psychic energy
(2) transformation of the biophysical energy of a human into psychic energy

This chapter is based on Freud's works [86] and the papers of Iurato and Khrennikov [115] and Iurato, Khrennikov, and Murtagh [116]. These two papers are devoted to mathematical modeling of unconscious dynamics and its connection with consciousness. This is one of the favorite topics of my research. The functioning of unconsciousness is represented as a classical dynamical system. Consciousness is represented as a system of observables transforming unconscious states into states of awareness. We remark that

Social Laser
Andrei Khrennikov

ISBN 978-981-4800-83-9 (Hardcover), 978-0-367-81720-6 (eBook)
www.jennystanford.com

the article of Iurato and Khrennikov [115] contains an extended bibliography which can be useful for the readers of this book.

I wrote a long series of papers and a few books on the mathematical modeling of Freud's theory of the interaction of unconscious and conscious mental states (see, e.g., the first chapters of monographs [128, 129, 145]). In these works, two classes of mathematical models were elaborated:

- p-adic (and more generally ultrametric) dynamical models of thinking
- quantum-like models

9.1 On the Notion of Representation According to Freud

9.1.1 The Three Levels or Orders of a Representation: Introduction

From a psychoanalytic viewpoint, Freud stated the first patterns of the human psyche taking into account models of physics, mainly adopting an energetic pattern drawn from hydrodynamics, and in particular choosing a diffusion scheme for the energy flow. This framework is clearly identifiable in the initial Freudian neurophysiological theories on human psychic mechanisms as exposed in the known *Project for a Scientific Psychology* of 1895 [86]. Since the first works till the final ones, as for example the celebrated yet unfinished compendium *An Outline of Psychoanalysis* of 1938 [86], Freud never left out his initial hydraulic model of psychic energy, ever trying to implement it with the other formulations worked out along the maturation and the evolution of his thought. Indeed, in this latest compendium, Freud gave a configuration of human psychic system based on the reciprocal and contemporary presence of three orders, namely the *emotional order*, the *thing representation order*, and the *word representation order*. In the Freudian oeuvre, these three orders provide the most elementary psychic act (*cathexis*) through object relationships.

Here, we briefly recall that a *cathexis* is a neologism introduced by English translators of Freud's works related to the German term *Besetzung.* It is a Freudian concept which would roughly refer to that charge of (psychic) energy associated with a thing, this last being meant in its widest sense (abstract or concrete thing, person, etc.). That is, cathexis is defined as the general process of investment or adhesion of mental or emotional energy to a person, object, representation, or idea. Such a term, like many other ones, has undergone a complex evolution along the whole Freudian thought in a way that makes it impossible to have a precise and unique meaning. In any case, its deepest roots should be retraced in the unconscious itself, where Freud even speaks of *unconscious cathexis* for the unrestrainable tendency of the unconscious to produce representation orders (like the ones we shall analyze below), that is to say, to refer to that intrinsic dynamicity of the unconscious (triggered by $\mathcal{UCC}$) which will give rise to every human intentional act (also in the sense of phenomenology), keep unchanged their presence although with probable role changing. They respectively rely on the *Triebrepräsentant* (or *Triebrepräsentanz*), that is, the drive or instinctual representation, on the *Sachvorstellung* (or *Objektvorstellung*), that is, the thing representation, and on the *Wortvorstellung*, that is, the word representation.

In any case, from a historical stance [86], Freud started from considering the notion of representation as meant by philosophy, that is to say, as the operation, or the act, with which consciousness reproduces an objective content. This is, broadly speaking, means: made by external objects or by internal psychic states, or else one of these objects themselves. Then, Freud considered a more specific psychological sense of this concept. To be precise, he intended to refer to the renewal of a perceptive experience in the absence of the sensorial stimulus, which should be meant as a knowledge act comprising the mere sensation and the next abstracting intellection. From this point onwards, Freud started to gradually consider three possible levels or orders of a representation understood in the latter terms.

9.1.2 On the First Representation Level or Order

The *representation of affect* can be defined as a medium for the affective expression, in a sense serving as its vehicle in mental dynamics. Freud clearly established [86] that instinct, situated, so to speak, at the "frontier" between body and mind, is expressed in mental dynamics by means of two main components: on the one hand, the *ideational representative*, and on the other, an energy charge, the *quota of affect* (*Affektbetrag*), whose fate can be distinct. This energy change, when it is temporarily without a representational support, can be qualitatively transformed into various emotions (e.g., fear, pleasure, anxiety) and, under certain conditions, cathexis ideational representations can await the arrival of such a charge that will embody and validate them. This is the case, for example, when anxiety "without an object" (that is, without a conscious object) becomes focused around a clearly identifiable phobogenic object. Thus, if we wish to make the idea of what representation of affect is, then we may metaphorically consider a traveling salesman "representing" a product, who has been delegated by his or her employer to pursue this function for the clients, in this analogy the instinct being the employer, and consciousness being the client. This last metaphor is coherent with the usage of the term *Repräsentanz* by Freud to which he intended to give the meaning of "delegate," "deputy," or "representative" just in the commercial sense.

Freud used the term *Repräsentanz* just to mean this function, that is to say, the representing or delegating function, and introduced it to explain that instinctual work imposed on the psyche because of its unavoidable and inseparable link with body. The drive or instinct coming from body is considered, in *Three Essays on the Theory of Sexuality* (1905, see [86]), as the *psychische Repräsentanz*, while, in *Instincts and Their Vicissitudes* (1915), this is considered as the *psychische Repräsentant*. The *psychische Repräsentanz* should be meant as psychic representative, while the *psychische Repräsentant* should be meant as psychic representation. Subsequently, from 1915 onward, Freud used only the latter, leaving out the former, but ever keeping a certain semantic ambiguity. In two metapsychological essays of 1915, namely *Die Verdrängung* [The Repression] (1915) and

Das Unbewusste [The Unconscious] (1915), Freud first claims [86] a clean distinction between an ideational element (representation) and an affective element (Affektbetrag). As claimed in [115, 116, 145], there is a control center CC, containing censorship agencies, which has functions of control, formulates problems and sends them to the unconscious domain $\mathcal{U}$. Among the many censorship functions played by CC, there is also that of *primal repression* which gives rise to the very early repressed material, the one lying into the deepest meanders of unconscious, the first hardcore, say $\mathcal{U}CC$ (see later), of the unconscious which is unexplorable in the psychoanalytic setting; it is placeable into the implicit memories and is inaccessible to the next explicit memories. This first core will run as an attractor of the other next repressed contents, coming from censorship functions which mainly work as repulsion centers, and mainly operating by means of *fixation* processes through which Affektbetrag attaches to ideational element (representation). Therefore, the concept of representation, in Freud, has undergone a complex evolution in which he paid attention first to the initial triad Triebrepräsentant-Sachvorstellung-Wortvorstellung, then to the pair Sachvorstellung-Wortvorstellung, until the last notion of *purposive idea* (or *finalized representation*), although all them are present more or less explicitly in the Freudian thought since its early steps.

The Triebrepräsentant is meant, by Freud, as the primary instinctual representative with which drives and endosomatic excitations find their psychic expressions. Sometimes, in the Freudian view, such a term refers to the ideational element, while other times it is wider and connotes affective elements. On the other hand, as recalled in [86], the instinctual representative (Triebrepräsentant), ideational representative, and psychical representative should be meant as synonymous of each other, to roughly mean some representative of the somatic with the psychic. In any case, Freud then pointed out a clear distinction between Affektbetrag and representation because both undergo a different fate. Hence he carried on with a deeper discussion on representation, distinguishing between representation of things and representation of words on the basis of his juvenile study on aphasia of 1891 which has tacitly played a fundamental role in the next development of the Freudian thought [86]. Nevertheless, also in this case, Freud seems not to depart much from the above semantic field identified by the instinctual representative (Triebrepräsentant), ideational representative, and psychical representative. Furthermore, as again pointed out in [86], even representation of things and representation of words might be considered as belonging to the same semantic field, so that it is allowed, eventually, to focus only on this last pair (thing representation/word representation), if nothing else, because Freud himself will later give a major load to it in the dawning of consciousness. However, along the main pathway comprehending the three representation orders Triebrepräsentant, Sachvorstellung, and Wortvorstellung, which we here suppose to be distinct, it is possible to consider the main link between corporal affect, representation of phantasmagoric and/or hallucinatory things of the scene of satisfaction, and/or dissatisfaction through which the affect's charge fulfils. Hence their symbolic representation by means of language[1] and through which secondary a process mainly performs, as one of the keystones of the psychic functioning according to Freud. But two crucial points of Freudian thought are that the instinctual charge, meant as an affective element, is independent of the representation, meant as an ideation element, and that representation of things and representation of words are tied together to give rise to a conscious representation belonging to the secondary process. The third representation order, therefore, bears in itself—roughly as in a (non-isolated) closed thermodynamical system—that received bodily energy amount (*Affektbetrag*) brought by the pair, roughly as in a (non-isolated) open thermodynamic system[2]—made by the first two representation orders tied together into the third one, that is to say, the secondary process, in the third representation order, makes bound (*bound energy*), through systems of symbols (among

---

[1] Nevertheless, there are other manners or ways of satisfaction not related with pure symbolic processes due to sublimation (and certain substitutive) phenomena, but linked to other degenerate phenomena like fetishism [115, 116].

[2] Thanks to the openness of this thermodynamical system, that is to say, thanks to the unavoidable great sensible perception of the human body to put itself in touch with the external world, it is possible to have the triggering of those irreversible processes (among which is the diffusion process of Triebrepräsentant) which will give rise to the human life pathway and its temporal evolution, and mainly due to those instability conditions of non-equilibrium given by the basic contrast, life instincts–death instincts) ruling unconscious dynamics by $\mathcal{UCC}$.

which are the linguistic ones) which replace or "stand for" things and affects, the open energy (*free energy*) freely arising from the primary process composed of the first two representation orders. This takes place similarly to a Hegelian *Aufheben*, that is to say, sticking at the thing representation affectively charged (around the cathexis object or thing)—that is, a signified—a chain of words as signifiers[3] (see also *associative chains*), which is possible thanks to verbalization system which mainly associates a mental image with a word.

The concept of affect had great importance in the Freudian framework since the first works of J. Breuer and Freud on hysteria whose symptoms are retraceable in a traumatic event without a suitable energy discharge (locked affect). These studies showed that affect may be not linked with representation. The problem was then treated deeply in the metapsychological studies *Die Verdrängung* [The Repression[4]] (1915) and *Das Unbewusste* [The Unconscious[5]] (1915). To be precise, in the first work, Freud states that

> we have dealt with the repression of an instinctual representative, and by the latter we have understood an idea or group of ideas which is cathexed with a definite quota of psychical energy (libido or interest) coming from an instinct. Clinical observation now obliges us to divide up what we have hitherto regarded as a single entity; for it shows us that besides the idea, some other element representing the instinct has to be taken into account, and that this other element undergoes vicissitudes of repression which may be quite different from those undergone by the idea. For this other element of the psychical representative the term *quota of affect* [*Affektbetrag*] has been generally adopted. It corresponds to the instinct in so far as the latter has become detached from the idea and finds expression, proportionate to its quantity, in processes which are sensed as affects. From this point on, in describing a case of repression, we shall have to follow up separately what, as the result of repression, becomes of the idea, and what becomes of the instinctual energy linked to it. ([86], Volume XIV, p. 152)

[3]We are very close to Lacan theory.
[4]See [86], Volume XIV, pp. 146–158.
[5]See [86], Volume XIV, pp. 159–216.

This first representation refers, therefore, to the basic *Affektbetrag*, that is to say, the primary emotional charge which is tied all around the object of cathexis. According to Freud, the affectivity concerns the bodily context and is defined as the subjective transposition of the quantity of instinctual energy. The body is broadly meant as the resonant place of emotions whose early roots are basically embodied; it is the early source of every human action and thought, as well as the primary means between internal and external world. The body enables the fundamental transformation of the quantitative kinesthetic, sense-perceptive, and somatic energy to other qualitative and quantitative forms of energy, including the qualitative one related to the emotional and rational thought. Freud considered a hydrodynamic model of corporal energy flow based on free diffusion processes of energy mainly regulated from pleasure and displeasure principles. Therefore, we must search out the early and primary origins of every human mental state in the corporal encompassing bodily energy providing the Triebrepräsentant. Broadly speaking, following Freud's original ideas, this bodily energy seems to be a physiological one meant in its truly quantitative nature and springing out from natural biochemical-physic processes which occur in our own body since the prenatal stage.[6] It will be just the Triebrepräsentant to convert part of this energetic amount into other qualitative forms of energy like the affective ones and those related to hysteretic and conversion phenomena. As indicated above, Freud adopted (also with J. Breuer) a hydraulic model for the energy flow questions variously involved in his framework, mainly based on the psychophysical and thermodynamical works of von Helmholtz, von Brücke, Rankine, and Thompson. Furthermore, it seems that Freud was also inspired by Herbart's psychological theory of mechanical representations (*Vorstellungmechanik*) in drawing up his theory of three-order psychic representations [86].

[6] Nevertheless, following the patterns of natural sciences, Freud didn't make hypotheses on the possible nature of the energy involved in the many issues raised along his studies and research. He considered energy in the same manner natural sciences treat energy questions.

9.1.3 On the Second and Third Representation Level or Order

As regards the other two representation orders, that is to say, the Sachvorstellung (i.e., the presentation of the thing, or thing-presentation) and the Wortvorstellung (i.e., the presentation of the word, or word-presentation), we first quote what Freud himself says in *Das Unbewusste* [The Unconscious].

> In schizophrenia words are subjected to the same process as that which makes the dream images out of latent dream-thoughts—to what we have called the primary psychical process. They undergo condensation, and by means of displacement transfer their cathexes to one another in their entirety. The process may go so far that a single word, if it is specially suitable on account of its numerous connections, takes over the representation of a whole train of thought. [. . .] If we ask ourselves what it is that gives the character of strangeness to the substitutive formation and the symptom in schizophrenia, we eventually come to realize that it is the predominance of what has to do with words over what has to do with things. [. . .] What has dictated the substitution is not the resemblance between the things denoted but the sameness of the words used to express them. Where the two—word and thing—do not coincide, the formation of substitutes in schizophrenia deviates from that in the transference neuroses. If now we put this finding alongside the hypothesis that in schizophrenia object-cathexes are given up, we shall be obliged to modify the hypothesis by adding that the cathexis of the word-presentations of objects is retained. What we have permissibly called the conscious presentation of the object can now be split up into the *presentation of the word* and the *presentation of the thing*; the latter consists in the cathexis, if not of the direct memory-images of the thing, at least of remoter memory-traces derived from these. We now seem to know all at once what the difference is between a conscious and an unconscious presentation. The two are not, as we supposed, different registrations of the same content in different psychical localities, nor yet different functional states of cathexis in the same locality; but the conscious presentation comprises the presentation of the thing plus the presentation of the word belonging to it, while the unconscious presentation is the presentation of the thing alone. The system *Ucs.* contains the

> thing cathexes of the objects, the first and true object-cathexes; the system *Pcs.* comes about by this thing-presentation being hypercathected through being linked with the word presentations corresponding to it. It is these hypercathexes, we may suppose, that bring about a higher psychical organization and make it possible for the primary process to be succeeded by the secondary process which is dominant in the *Pcs.* Now, too, we are in a position to state precisely what it is that repression denies to the rejected presentation in the transference neuroses: what it denies to the presentation is translation into words which shall remain attached to the object. A presentation which is not put into words, or a psychical act which is not hypercathected, remains thereafter in the *Ucs.* in a state of repression. ([86], Volume XIV, pp. 200–201)

Therefore, the second representation order, namely the Sachvorstellung, gives rise the eidetic-perceptive representations of external objects, to gestalt effects, spatial-temporally coordinated among them in a varied manner, while the third representation order, namely the Wortvorstellung, provides the linguistic-symbolic representation. In the first order prevails, so to speak, a quantitative-diffusive logic due to bodily energy encompassment, in the second order a qualitative-associative or combinatory logic due to the cooperative action of different sensory inputs and images, and in the third order a set of formal logics related with the previous ones. The intertwinement between these three orders to give rise explicitly (as a consciousness act) to the third one, in which the first two are implicitly present and tied together, is the basis of the Freudian nonpathological running of consciousness. This interlacement is also placeable into the above framework considering the relationships conscious-unconscious $\mathcal{C}-\mathcal{U}$ because the first two representation orders mainly are in the unconscious domain and arrange the primary process, while the third one is in the conscious domain but is closely related with the first two.

Chapter 10

Introduction to Quantum Theory

This chapter contains a brief introduction to mathematical formalism and QM axiomatics. It is oriented to non-physicists. Since *QM is a statistical theory* it is natural to start with the classical probability model (Kolmogorov [191], 1933). Then we present the basics of the theory of Hilbert spaces and Hermitian operators and the representation of pure and mixed states by normalized vectors and density operators. This introduction is sufficient to formulate the axiomatics of QM in the form of five postulates. The projection postulate (the most questionable postulate of QM) is presented in a separate section. We sharply distinguish between the cases of quantum observables represented by Hermitian operators with nondegenerate and degenerate spectra, von Neumann's and Lüders's forms of the projection postulate. The axiomatics is completed by a short section on the main interpretations of QM. The projection postulate (Lüders's form) plays a crucial role in the definition of quantum conditional (transition) probability. By operating with the latter we consider interference of probabilities for two incompatible observables as a modification of the formula of total probability by adding the interference term. This viewpoint to interference of probabilities was elaborated in a series of works by Khrennikov. Since classical probability theory is based on the

Social Laser
Andrei Khrennikov

ISBN 978-981-4800-83-9 (Hardcover), 978-0-367-81720-6 (eBook)
www.jennystanford.com

Boolean algebra of events, a violation of the law of total probability can be treated as the probabilistic sign of a violation of the laws of the Boolean logics. From this viewpoint, quantum theory can be considered as representing a new kind of logic, so-called quantum logic. The latter is also briefly presented in a separate section. We continue this review with a new portion of quantum mathematics, namely the notion of the tensor product of Hilbert spaces and the tensor product of operators. After the section on Dirac's notation with ket and bra vector, we discuss briefly the notion of qubit and entanglement of a few qubits. This chapter concludes with a presentation of the detailed analysis of the probabilistic structure of the two-slit experiment in a bunch of different experimental contexts. This contextual structure leads to a violation of the law of total probability and non-Kolmogorovean probabilistic structure of this experiment.

We hope that this chapter would be interesting for newcomers to quantum-like modeling. Maybe even experts can find something useful, say the treatment of interference of probabilities as a violation of the law of total probability.

10.1 Classical Probability Theory: Kolmogorov's Measure-Theoretic Axiomatics

The main aim of QM is to provide probabilistic predictions on the results of measurements. Moreover, statistics of the results of measurements of a single quantum observable can be described by classical probability theory. We now present an elementary introduction to this theory.

The *Kolmogorov probability space* [191] is any triple

$$(\Omega, \mathcal{F}, \mathbf{P}),$$

where Ω is a set of any origin and $\mathcal{F}$ is a σ-algebra of its subsets and $\mathbf{P}$ is a probability measure on $\mathcal{F}$. The set Ω represents random parameters of the model. Kolmogorov called elements of Ω *elementary events*. This terminology is standard in mathematical literature. Sets of elementary events are regarded as *events*. The key point of Kolmogorov's axiomatization of probability theory is that not any subset of Ω can be treated as an event. For any stochastic model, the system of events $\mathcal{F}$ is selected from the very beginning. The key mathematical point is that $\mathcal{F}$ has to be a σ *algebra.* (Otherwise it would be very difficult to develop a proper notion of integral. And the latter is needed to define average of a random variable.)

We remind that a σ algebra is a system of sets which contains Ω and an empty set, and it is closed with respect to the operations of countable union and intersection and to the operation of taking the complement of a set. For example, the collection of all subsets of Ω is σ algebra. This σ algebra is used in the case of finite or countable sets:

$$\Omega = \{\omega_1, \ldots, \omega_n, \ldots\}. \tag{10.1}$$

However, for "continuous sets," e.g., $\Omega = [a, b] \subset \mathbf{R}$, the collection of all possible subsets is too large to have applications. Typically it is impossible to describe a σ algebra in the direct terms. To define a σ algebra, one starts with a simple system of subsets of Ω and then considers the σ algebra which is generated from this simple system with the aid of aforementioned operations. In particular, one of the most important for applications σ algebras, the so-called *Borel σ lgebra*, is constructed in this way by staring with the system consisting of all open and closed intervals of the real line. In a metric space (in particular, a Hilbert space), the Borel σ algebra is constructed by starting with the system of all open and closed balls.

Finally, we remark that in American literature the term σ field is typically used, instead of σ algebra.

The probability is defined as a *measure*, i.e., a map from $\mathcal{F}$ to nonnegative real numbers which is σ-additive:

$$\mathbf{P}(\cup_j A_j) = \sum_j \mathbf{P}(A_j), \tag{10.2}$$

where $A_j \in \mathcal{F}$ and $A_i \cap A_j = \emptyset, i \neq j$. The probability measure is always normalized by one:

$$\mathbf{P}(\Omega) = 1. \tag{10.3}$$

In the case of a discrete probability space (see (10.1)), the probability measures have the form

$$\mathbf{P}(A) = \sum_{\omega_j \in A} p_j, \quad p_j = \mathbf{P}(\{\omega_j\}).$$

In fact, any finite measure μ, i.e., $\mu(\Omega) < \infty$, can be transformed into the probability measure by normalization:

$$\mathbf{P}(A) = \frac{\mu(A)}{\mu(\Omega)}, \; A \in \mathcal{F}. \tag{10.4}$$

A (real) random variable is a map $\xi : \Omega \to \mathbf{R}$ which is measurable with respect to the Borel σ algebra $\mathcal{B}$ of $\mathbf{R}$ and the σ algebra $\mathcal{F}$ of Ω. The latter means that, for any set $B \in \mathcal{B}$, its preimage $\xi^{-1}(B) = \{\omega \in \Omega : \xi(\omega) \in B\}$ belongs to $\mathcal{F}$. This condition provides the possibility of assigning the probability to the events of the type "values of ξ belong to a (Borel) subset of the real line." The probability distribution of ξ is defined as

$$\mathbf{P}_\xi(B) = \mathbf{P}(\xi^{-1}(B)). \tag{10.5}$$

In the same way we define the real (and complex) vector valued random variables, $\xi : \Omega \to \mathbf{R}^n$ ($\xi : \Omega \to \mathbf{C}^n$).

A random variable is called discrete if its image consists of finite or countable number of points, $\xi = \alpha_1, \ldots, \alpha_n, \ldots$. In this case its probability distribution has the form

$$\mathbf{P}(B) = \sum_{\alpha_j \in B} P_{\alpha_j}, \; P_{\alpha_j} = \mathbf{P}(\omega \in \Omega : \xi(\omega) = \alpha_j). \tag{10.6}$$

The mean value (average) of a real valued random variable is defined as its integral (the Lebesgue integral)

$$E\xi = \int_\Omega \xi(\omega) dP(\omega). \tag{10.7}$$

For a discrete random variable, its mean value has the simple representation

$$E\xi = \sum_{\alpha_j \in B} \alpha_j P_{\alpha_j}. \tag{10.8}$$

In the Kolmogorov model the conditional probability is *defined* by the *Bayes formula*

$$\mathbf{P}(B|A) = \frac{\mathbf{P}(B \cap A)}{\mathbf{P}(A)}, \; \mathbf{P}(A) > 0. \tag{10.9}$$

We stress that other axioms are independent of this definition.

We also present the *formula of total probability* (FTP), which is a simple consequence of the Bayes formula. Consider the pair a and b of discrete random variables. Then

$$\mathbf{P}(b = \beta) = \sum_\alpha \mathbf{P}(a = \alpha)\mathbf{P}(b = \beta | a = \alpha). \tag{10.10}$$

Thus the b probability distribution can be calculated from the a probability distribution and the conditional probabilities $\mathbf{P}(b = \beta | a = \alpha)$. These conditional probabilities are known as *transition probabilities.*

10.2 Mathematical Structure of Quantum Theory

10.2.1 Complex Hilbert Space

We recall the definition of a complex Hilbert space. Denote it by H. This is a complex linear space endowed with a scalar product (a positive-definite nondegenerate Hermitian form) which is complete with respect to the norm corresponding to the scalar product $\langle \cdot | \cdot \rangle$. The norm is defined as

$$\|\phi\| = \sqrt{\langle \phi | \phi \rangle}.$$

In the finite-dimensional case the norm and, hence, completeness are of no use. Thus those who have no idea about functional analysis (but know the essentials of linear algebra) can treat H simply as a finite-dimensional complex linear space with the scalar product.

For a complex number $z = x + iy$, $x, y \in \mathbf{R}$, its conjugate is denoted by $\bar{z}$, where $\bar{z} = x - iy$. The absolute value of z is given by $|z|^2 = z\bar{z} = x^2 + y^2$.

For the reader's convenience, we recall that the scalar product is a function from the Cartesian product $H \times H$ to the field of complex numbers $\mathbb{C}$, $\psi_1, \psi_2 \to \langle \psi_1 | \psi_2 \rangle$, having the following properties:

(1) Positive-definiteness: $\langle \psi | \psi \rangle \geq 0$ with $\langle \psi, \psi \rangle = 0$ if and only if $\psi = 0$.
(2) Conjugate symmetry: $\langle \psi_1 | \psi_2 \rangle = \overline{\langle \psi_2 | \psi_1 \rangle}$
(3) Linearity with respect to the second argument[1]: $\langle \phi | k_1 \psi_1 + k_2 \psi_2 \rangle = k_1 \langle \phi | \psi_1 \rangle + k_2 \langle \phi | \psi_2 \rangle$, where k_1 and k_2 are complex numbers.

It generates the norm on $H : \|\psi\| = \sqrt{\langle \psi | \psi \rangle}$.

[1]In mathematical texts one typically considers linearity with respect to the first argument. Thus a mathematician has to pay attention to this difference.

A reader who does not feel comfortable in the abstract framework of functional analysis can simply proceed with the Hilbert space $H = \mathbf{C}^n$, where $\mathbf{C}$ is the set of complex numbers, and the scalar product

$$\langle u|v\rangle = \sum_i u_i \bar{v}_i,\ u = (u_1, \ldots, u_n),\ v = (v_1, \ldots, v_n). \qquad (10.11)$$

In this case the above properties of a scalar product can be easily derived from Eq. 10.11. Instead of linear operators, one can consider matrices.

10.2.2 Linear Operators

We also recall a few basic notions of the theory of linear operators in a complex Hilbert space. A map $\widehat{a} : H \to H$ is called a linear operator if it maps a linear combination of vectors into a linear combination of their images:

$$\widehat{a}(\lambda_1\psi_1 + \lambda_2\psi_2) = \lambda_1\widehat{a}\psi_1 + \lambda_2\widehat{a}\psi_2,$$

where $\lambda_j \in \mathbf{C}$, $\psi_j \in H$, $j = 1, 2$.

For a linear operator $\widehat{a}$ the symbol $\widehat{a}^*$ denotes its *adjoint operator* which is defined by the equality

$$\langle \widehat{a}\psi_1|\psi_2\rangle = \langle \psi_1|\widehat{a}^*\psi_2\rangle. \qquad (10.12)$$

Let us select in H some orthonormal basis (e_i), i.e., $\langle e_i|e_j\rangle = \delta_{ij}$. By denoting the matrix elements of the operators $\widehat{a}$ and $\widehat{a}^*$ as a_{ij} and a^*_{ij}, respectively, we rewrite the definition (10.12) in terms of the matrix elements:

$$a^*_{ij} = \bar{a}_{ji}.$$

A linear operator $\widehat{a}$ is called *Hermitian* if it coincides with its adjoint operator:

$$\widehat{a} = \widehat{a}^\star.$$

If an orthonormal basis in H is fixed, (e_i), and $\widehat{a}$ is represented by its matrix, $A = (a_{ij})$, where $a_{ij} = \langle \widehat{a}e_i|e_j\rangle$, then it is Hermitian if and only if

$$\bar{a}_{ij} = a_{ji}.$$

We remark that, for a Hermitian operator, all its eigenvalues are real. In fact, this was one of the main reasons to represent quantum observables by Hermitian operators. In quantum formalism, the spectrum of a linear operator (the set of eigenvalues while we are in the finite dimensional case) coincides with the set of possibly observable values (section 10.3, Postulate 3). We also recall that the eigenvectors of Hermitian operators corresponding to different eigenvalues are orthogonal. This property of Hermitian operators plays some role in the justification of the projection postulate of QM (see section 10.3.1).

A linear operator $\widehat{a}$ is *positive-semidefinite* if, for any $\phi \in H$,

$$\langle \widehat{a}\phi|\phi\rangle \geq 0.$$

This is equivalent to the positive-semidefiniteness of its matrix.

For a linear operator $\widehat{a}$, its trace is defined as the sum of the diagonal elements of its matrix in any orthonormal basis:

$$\mathrm{Tr}\,\widehat{a} = \sum_i a_{ii} = \sum_i \langle \widehat{a}e_i|e_i\rangle,$$

i.e., this quantity does not depend on a basis.

Let L be a subspace of H. The orthogonal projector $P : H \to L$ onto this subspace is a Hermitian, idempotent (i.e., coinciding with its square) and positive-semidefinite operator[2]:

(1) $P^* = P$;
(2) $P^2 = P$;
(3) $P \geq 0$.

Here (3) is a consequence of (1) and (2). Moreover, an arbitrary linear operator satisfying (1) and (2) is an orthogonal projector—onto the subspace PH.

10.2.3 Representation of (Pure) States by Normalized Vectors

Pure quantum states are represented by normalized vectors, $\psi \in H : \|\psi\| = 1$. Two colinear vectors,

$$\psi' = \lambda\psi,\ \lambda \in \mathbf{C},\ |\lambda| = 1, \tag{10.13}$$

[2]To simplify formulas, we shall not put the operator label "hat" in the symbols denoting projectors, i.e., $P \equiv \widehat{P}$.

represent the same pure state. Thus, rigorously speaking, a pure state is an equivalence class of vectors having the unit norm $\psi' \sim \psi$ for vectors coupled by (10.13). The unit sphere of H is split into disjoint classes: pure states. However, in concrete calculations, one typically uses just concrete representatives of equivalent classes, i.e., one works with normalized vectors.

Each pure state can also be represented as the projection operator P_ψ which projects H onto a one-dimensional subspace based on ψ. For a vector $\phi \in H$,

$$P_\psi \phi = \langle \phi | \psi \rangle \, \psi. \tag{10.14}$$

The trace of the one-dimensional projector P_ψ equals 1:

$$\text{Tr } P_\psi = \langle \psi | \psi \rangle = 1.$$

We summarize the properties of the operator P_ψ representing the pure state ψ. It is

(1) Hermitian,
(2) positive-semidefinite,
(3) trace one,
(4) idempotent.

Moreover, any operator satisfying (1)–(4) represents a pure state. Properties (1) and (4) characterize orthogonal projectors, and property (2) is their consequence. Property (3) implies that the projector is one-dimensional.

10.2.4 Representation of Mixed States by Density Operators

The next step in the development of QM was the extension of the class of quantum states, from pure states represented by one-dimensional projectors to states represented by linear operators having properties (1)–(3). Such operators are called *density operators.* (This nontrivial step of extension of the class of quantum states was based on the efforts of Landau and von Neumann.) The symbol $D(H)$ denotes the space of density operators in the complex Hilbert space H.

One typically distinguishes pure states, as represented by one-dimensional projectors, and mixed states, the density operators which cannot be represented by one-dimensional projectors. The term *mixed* has the following origin: any density operator can be represented as a *mixture* of pure states (ψ_i):

$$\rho = \sum_i p_i P_{\psi_i}, \quad p_i \in [0, 1], \sum_i p_i = 1. \tag{10.15}$$

(To simplify formulas, we shall not put the operator-label "hat" in the symbols denoting density operators, i.e., $\rho \equiv \widehat{\rho}$.) The state is pure if and only if such a mixture is trivial: all p_i, besides one, equal zero. However, by operating with the term *mixed state* one has to take into account that the representation in the form (10.15) is not unique. The same mixed state can be presented as mixtures of different collections of pure states.

Any operator ρ satisfying (1)–(3) is diagonalizable (even in infinite-dimensional Hilbert space), i.e., in some orthonormal basis it is represented as a diagonal matrix, $\rho = \text{diag}(p_j)$, where $p_j \in [0, 1], \sum_j p_j = 1$. Thus it can be represented in the form (10.15) with mutually orthogonal one-dimensional projectors. Property (4) can be used to check whether a state is pure or not. We point out that pure states are merely mathematical abstractions; in real experimental situations it is possible to prepare only mixed states. One defines the degree of *purity* as $\text{Tr}\rho^2$. Experimenters are satisfied by getting this quantity near one.

10.2.5 Hilbert Space of Square Integrable Functions

Although we generally proceed with finite-dimensional Hilbert spaces, it is useful to mention the most important example of infinite-dimensional Hilbert space used in QM. Consider the space of complex-valued functions, $\psi : \mathbb{R}^m \to \mathbb{C}$, which are square integrable with respect to the Lebesgue measure on $\mathbb{R}^m$:

$$\|\psi\|^2 = \int_{\mathbb{R}^m} |\psi(x)|^2 dx < \infty. \tag{10.16}$$

It is denoted by the symbol $L^2(\mathbb{R}^m)$. Here the scalar product is given by

$$\langle \psi_1 | \psi_2 \rangle = \int_{\mathbb{R}^m} \bar{\psi}_1(x) \psi_2(x) dx.$$

A delicate point is that, for some measurable functions, $\psi : \mathbb{R}^m \to \mathbb{C}$, which are not identically zero, the integral

$$\int_{\mathbb{R}^m} |\psi(x)|^2 dx = 0. \tag{10.17}$$

We remark that the latter equality implies that $\psi(x) = 0$ a.e. (almost everywhere). Thus the quantity defined by Eq. 10.16 is, in fact, not the norm: $\|\psi\| = 0$ does not imply that $\psi = 0$. To define a proper Hilbert space, one has to consider as its elements not simply functions, but classes of equivalent functions, where the equivalence relation is defined as $\psi \sim \phi$ if and only if $\psi(x) = \phi(x)$ a.e. In particular, all functions satisfying Eq. 10.17 are equivalent to the zero function.

10.3 Postulates of Quantum Mechanics

We state again that H denotes a complex Hilbert space with the scalar product $\langle \cdot, \cdot \rangle$ and the norm $\| \cdot \|$ corresponding to the scalar product.

Postulate 1. (The mathematical description of quantum states) *Quantum (pure) states (wave functions) are represented by normalized vectors ψ (i.e., $\|\psi\|^2 = \langle \psi, \psi \rangle = 1$) of a complex Hilbert space $\mathcal{H}$. Every normalized vector $\psi \in H$ may represent a quantum state. If a vector ψ corresponding to a state is multiplied by any complex number c, $|c| = 1$, the resulting vector will correspond to the same state.*[3]

The physical meaning of "a quantum state" is not defined by this postulate (see section 10.5).

Postulate 2. (The mathematical description of physical observables) *A physical observable a is represented by a Hermitian operator $\widehat{a}$ in complex Hilbert space H. Different observables are represented by different operators.*

Postulate 3. (Spectral) *For a physical observable a which is represented by the Hermitian operator $\widehat{a}$ we can predict (together with some probabilities) values $\lambda \in$* $\mathrm{Spec}(\widehat{a})$.

[3]Thus, states are given by elements of the unit sphere of the Hilbert space H.

We restrict our considerations by the simplest Hermitian operators which are analogous to discrete random variables in classical probability theory. We recall that a Hermitian operator $\widehat{a}$ has a *purely discrete spectrum* if it can be represented as

$$\widehat{a} = \alpha_1 P^a_{\alpha_1} + \ldots + \alpha_m P^a_{\alpha_m} + \ldots, \quad \alpha_m \in \mathbf{R}, \tag{10.18}$$

where $P^a_{\alpha_m}$ are orthogonal projection operators related to the orthonormal eigenvectors $\{e^a_{km}\}_k$ of $\widehat{a}$ corresponding to the eigenvalues α_m by

$$P^a_{\alpha_m}\psi = \sum_k \langle \psi | e^a_{km} \rangle e^a_{km}, \quad \psi \in H. \tag{10.19}$$

Here k labels the eigenvectors e^a_{km} which belong to the same eigenvalue α_m of $\widehat{a}$.

Postulate 4. (Born's rule) *Let a physical observable a be represented by a Hermitian operator $\widehat{a}$ with a purely discrete spectrum. The probability $\mathbf{P}_\psi(a = \alpha_m)$ to obtain the eigenvalue α_m of $\widehat{a}$ for measurement of a in a state ψ is given by*

$$\mathbf{P}_\psi(a = \alpha_m) = \| P^a_m \psi \|^2. \tag{10.20}$$

If the operator $\widehat{a}$ has nondegenerate (purely discrete) spectrum, then each α_m is associated with one-dimensional subspace. The latter can be fixed by selecting any normalized vector, say e^a_m. In this case orthogonal projectors act simply as

$$P^a_{\alpha_m}\psi = \langle \psi | e^a_m \rangle e^a_m. \tag{10.21}$$

The formula (10.20) takes a simple form:

$$\mathbf{P}_\psi(a = \alpha_m) = |\langle \psi | e^a_m \rangle|^2. \tag{10.22}$$

It is Born's rule in the Hilbert space formalism.

To obtain original Born's rule, one should choose H as the L_2 space of square integrable functions, $\psi : \mathbf{R} \mapsto \mathbf{C}$. (We consider a one-dimensional particle). The position observable x is represented by the multiplication operator $\widehat{x}$

$$\widehat{x}(\psi)(x) = x\psi(x). \tag{10.23}$$

This operator has a continuous spectrum. It coincides with the whole real line. So, this operator is unbounded. Its eigenvectors do not belong to the L_2 space. They are given by Dirac's δ functions, i.e., these are generalized eigenfunctions (see Dirac [65])

$$\widehat{x}(e_\alpha)(x) = \alpha e_\alpha(x), \ \ \alpha \in \mathbf{R}, \tag{10.24}$$

where $e_\alpha(x) = \delta(x - \alpha)$. One can reasonably define paring:

$$\langle \psi | e_\alpha \rangle = \psi(\alpha). \tag{10.25}$$

Then the rule (10.22) gives

$$\mathbf{P}_\psi(x = \alpha) = |\psi(\alpha)|^2. \tag{10.26}$$

This rule was invented in the following way. Originally Schrödinger considered the ψ function as a classical field. The quantity $E(\alpha) = |\psi(\alpha)|^2$ was interpreted as the density of the electron's charge. Born invented the rule (10.26) by criticizing Schrödinger's interpretation. Instead of the charge's density, he considered this quantity as the probability density. The latter induces automatically the normalization condition

$$1 = \int_{-\infty}^{+\infty} |\psi(\alpha)|^2 = \langle \psi | \psi \rangle$$

which was absent in Schrödinger's model. After a few years of struggle for the classical field interpretation of the wave function, Schrödinger gave up and he kept to Born's probabilistic interpretation.

In the same way one can consider the *momentum observable*. Schrödinger defined the momentum operator as

$$\widehat{p}(\psi)(x) = -i\frac{d}{dx}\psi(x). \tag{10.27}$$

(We eliminate the Planck constant from consideration by choosing a corresponding system of units). It is easy to see that its spectrum is also continuous and it coincides with **R**. Its generalized eigenfunctions can be easily found from the equation

$$-i\frac{d}{dx}f_\beta(x) = \beta f_\beta(x), \ \ \beta \in \mathbf{R}.$$

Thus $f_\beta(x) = e^{i\beta x}$. Thus by (10.22)

$$\mathbf{P}_\psi(p = \beta) = |\langle \psi | f_\beta \rangle|^2. \tag{10.28}$$

By taking into account that

$$\langle \psi | f_\beta \rangle = \int_{-\infty}^{+\infty} \psi(x) e^{-i\beta x} dx = \tilde{\psi}(\beta)$$

is the Fourier transform of ψ, we write Born's rule for the momentum measurement as

$$\mathbf{P}_\psi(p = \beta) = |\tilde{\psi}(\beta)|^2, \tag{10.29}$$

cf. (10.26).

We point out that, for any fixed quantum state ψ, each quantum observable $\widehat{a}$ can be represented as a classical random variable (section 10.1). In the discrete case the corresponding probability distribution is defined as

$$\mathbf{P}(A) = \sum_{\alpha_m \in A} \mathbf{P}_\psi(a = \alpha_m),$$

where $\mathbf{P}_\psi(a = \alpha_m)$ is given by Born's rule.

Thus, each concrete quantum measurement can be described by the classical probability model.

Problems (including deep interpretational issues) arise only when one tries to classically describe the data collected for a few incompatible observables.

By using Born's rule (10.20) and the classical probability definition of average (section 10.1), it is easy to see that the average value of the observable a in the state ψ (belonging to the domain of definition of the corresponding operator $\widehat{a}$) is given by

$$\langle a \rangle_\psi = \langle \widehat{a}\, \psi | \psi \rangle. \tag{10.30}$$

For example, for an observable represented by an operator with the purely discrete spectrum, we have

$$\langle a \rangle_\psi = \sum_m \alpha_m \mathbf{P}_\psi(a = \alpha_m) = \sum_m \alpha_m \| P_m^a \psi \|^2 = \langle \widehat{a}\, \psi | \psi \rangle.$$

Postulate 5. (Time evolution of wave function) *Let $\widehat{\mathcal{H}}$ be the Hamiltonian of a quantum system, i.e., the Hermitian operator corresponding to the energy observable. The time evolution of the wave function $\psi \in H$ is described by the Schödinger equation*

$$i \frac{d}{dt} \psi(t) = \widehat{\mathcal{H}} \psi(t) \tag{10.31}$$

with the initial condition $\psi(0) = \psi$.

10.3.1 Projection Postulate, von Neumann versus Lüders

The projection postulate is one of the most questionable and debatable postulates of QM. We present it in a separate section to distinguish it from other postulates of QM, Postulates 1–5, which are commonly accepted.

Consider pure state ψ and quantum observable (Hermitian operator) $\widehat{a}$ representing some physical observable a. Suppose that $\widehat{a}$ has a nondegenerate spectrum; denote its eigenvalues by $\alpha_1, \ldots, \alpha_m, \ldots$ and the corresponding eigenvectors by $e_1^a, \ldots, e_m^a, \ldots$ (here $\alpha_i \neq \alpha_j$, $i \neq j$.) This is an orthonormal basis in H. We expand the vector ψ with respect to this basis:

$$\psi = k_1 e_1^a + \ldots + k_m e_m^a + \ldots, \tag{10.32}$$

where (k_j) are complex numbers such that

$$\|\psi\|^2 = |k_1|^2 + \ldots + |k_m|^2 + \ldots = 1. \tag{10.33}$$

By using the terminology of linear algebra we say that the pure state ψ is a *superposition* of the pure states e_j. The von Neumann projection postulate describes the post-measurement state and it can be formulated as follows:

Postulate 6VN. (Projection postulate, von Neumann) *Measurement of observable a (represented by Hermitian operator $\widehat{a}$) resulting in output α_i induces reduction in superposition (10.32) to the basis vector e_i^a.*

The procedure described by the projection postulate can be interpreted in the following way:

> Superposition (10.32) reflects uncertainty in the results of measurements for an observable a. Before measurement a quantum system "does not know how it will answer to the question a." Born's rule presents potentialities for different answers. Thus a quantum system in the superposition state ψ does not have propensity to any value of a as its objective property. After the measurement the superposition is reduced to the single term in the expansion (10.32) corresponding to the value of a obtained in the process of measurement.

Now consider an arbitrary quantum observable a represented by Hermitian operator $\widehat{a}$ with a purely discrete spectrum, i.e., $\widehat{a} =$

$\alpha_1 P^a_{\alpha_1} + \ldots + \alpha_m P^a_{\alpha_m} + \ldots$. The Lüders projection postulate describes the post-measurement state and it can be formulated as follows:

Postulate 6L. (Projection postulate, Lüders) *Measurement of observable a resulting in output α_m induces projection of state ψ on state*

$$\psi_{\alpha_m} = \frac{P^a_{\alpha_m}\psi}{\|P^a_{\alpha_m}\psi\|}.$$

In contrast to the majority of books on quantum theory, we sharply distinguish the cases of quantum observables with nondegenerate and degenerate spectra. Von Neumann formulated Postulate 6VN only for observables with nondegenerate spectra. Lüders "generalized" von Neumann's postulate to the case of observables with degenerate spectra. However, for such observables, von Neumann formulated [235] a postulate which is different from Postulate 6L. The post-measurement state need not be again a pure state; it can be a mixed state as well.

10.4 Operator Quantization: From Functions on Classical Phase Space to Hermitian Operators

We remark that the operators of position and momentum, $\widehat{x}$ and $\widehat{p}$ (see Eqs. 10.23 and 10.27) do not commute and they satisfy Heisenberg's *canonical commutation relation*

$$[\widehat{x}, \widehat{p}] = i. \tag{10.34}$$

Consider any real-valued function on the classical phase space, i.e., a function of classical coordinate and momentum, $f(x, p)$. The quantization procedure is the map

$$f \mapsto \widehat{f} = f(\widehat{x}, \widehat{p}). \tag{10.35}$$

In general, it is a tricky mathematical problem to define a function of two noncommuting operators. It is typically done by using the calculus of pseudo-differential operators. However, in the simplest case the operator of energy $\widehat{\mathcal{H}}$ can be easily defined. Consider a classical particle with the mass m moving in the potential $V(x)$. Its Hamiltonian function (representing classical energy of this particle) is given by

$$\mathcal{H}(x, p) = \frac{p^2}{2m} + V(x). \tag{10.36}$$

Quantization gives us the operator

$$\widehat{\mathcal{H}} = \mathcal{H}(\widehat{x}, \widehat{p}) = \frac{\widehat{p}^2}{2m} + V(x). \tag{10.37}$$

10.5 Two Basic Interpretations of a Quantum State

Now we are going to discuss one of the most important and complicated issues of quantum foundations, the problem of an interpretation of a quantum state. There were elaborated numerous interpretations which can differ crucially from each other (see [173]). This huge diversity of interpretations is a sign of the deep crises in quantum foundations. We present the two basic interpretations.

The ensemble interpretation. *A (pure) quantum state provides a description of certain statistical properties of an ensemble of similarly prepared quantum systems.*

This interpretation is upheld, for example, by Einstein, Popper, Blokhintsev, Margenau, Ballentine, Klyshko, and in recent years by, for example, De Muynck, De Baere, Holevo, Santos, Khrennikov, Nieuwenhuizen, Adenier, Groessing, and many others.

The Copenhagen interpretation. *A (pure) quantum state provides the complete description of an individual quantum system.*

This interpretation was supported by a great variety of members, from Schrödinger's original attempt to identify the electron with a wave function solution of his equation to the several versions of the Copenhagen interpretation (e.g., Heisenberg, Bohr, Pauli,

Dirac, von Neumann, Landau, Fock, and recent years by, e.g., Greenberger, Mermin, Lahti, Peres, and Summhammer). Nowadays the individual interpretation is extremely popular, especially in quantum information and computing.

Andrei Grib (who participated in the seminars of Vladimir A. Fock in Leningrad) told me about the correspondence between Bohr and Fock on the individual interpretation. Fock was the main supporter of this interpretation. Although Bohr himself had doubts about its consistency, Fock demonstrated to Bohr inconsistency in the Einsteinian ensemble interpretation and convinced him to use the individual interpretation.

Instead of Einstein's term *ensemble interpretation*, L. Ballentine [17–19] used the term *statistical interpretation.* However, Ballentine's term is rather misleading, because the term *statistical interpretation* was also used by J. von Neumann [235] for individual randomness! For him *statistical interpretation* had a meaning totally different from the Ballentine's *ensemble-statistical interpretation.* J. von Neumann wanted to emphasize the difference between deterministic (Newtonian) classical mechanics in that the state of a system is determined by values of two observables (position and momentum) and quantum mechanics in that the state is determined not by values of observables, but by probabilities. We shall follow Albert Einstein and use the term *ensemble interpretation.*

We remark that following von Neumann [235] the supporters of the individual interpretation believe in irreducible quantum randomness, i.e., that the behavior of an individual quantum system is irreducibly random. Why does it behave in such a way? Because it is quantum, it can behave so unusually. Nowadays this von Neumann claim is used to justify the superiority of quantum technology over classical technology—for example, the superiority of quantum random generators.

We also remark that the notion "Copenhagen interpretation" is quite ambiguous. There exists a bunch of interpretations in the spirit of Copenhagen (here we follow Plotnitsky [205–208]; see also Plotnitsky and Khrennikov [209]). The interpretation presented in this section is due to von Neumann [235].

We remark that the supporters of the ensemble interpretation are sharply split into two camps:

(1) those who believe in hidden variables
(2) those who believe in no-go theorems (e.g., Bell's theorem) and reject the possibility of introducing physically natural hidden variables

From my viewpoint the position of believers in the possibility to go beyond QM is more consistent than the position of unbelievers. For believers, quantum randomness is usual classical probabilistic randomness, but at the level of hidden variables. If one uses the statistical interpretation and believes in the common interpretation of violation of Bell-type inequalities, he is in a difficult situation. It is not easy to justify the statistical interpretation in the absence of hidden variables. Bell resolved this problem by considering nonlocal theories with hidden variables in the spirit of Bohmian mechanics. I think that if one accepts consequences of the no-go theorems, it would be more natural to use the individual interpretation, in the spirit of Copenhagen.

Maybe the reader has already understood that debates on the interplay of possible interpretations is very similar to the old days religious debates. This interplay brings chaos in thinking. For example, nowadays the majority of researchers working in quantum information believe both in von Neumann's irreducible quantum randomness and in Bell's nonlocality. Von Neumann by himself considered QM as a local theory. Bell by himself invented nonlocality in quantum foundations just to proceed with the ensemble interpretation and without von Neumann's irreducible randomness.

10.6 Conditional Probability in Quantum Formalism

In the classical Kolmogorov probabilistic model (section 10.1), besides probabilities one operates with the conditional probabilities defined by the *Bayes formula* (see section 10.1, formula (10.9)).

Born's postulate defining quantum probability should also be completed by a definition of conditional probability. We have remarked that, for one concrete observable, the probability given by Born's rule can be treated classically. However, the definition of the conditional probability involves two observables. Such situations cannot be treated classically (at least straightforwardly, cf. section 10.14.1). Thus, conditional probability is really quantum probability.

Let physical observables a and b be represented by Hermitian operators with purely discrete (may be degenerate) spectra:

$$\widehat{a} = \sum_m \alpha_m P^a_{\alpha_m}, \ \widehat{b} = \sum_m \beta_m P^b_{\beta_m} \tag{10.38}$$

Let ψ be a pure state and let $P^a_{\alpha_k}\psi \neq 0$. Then the probability to get the value $b = \beta_m$ under the condition that the value $a = \alpha_k$ was observed in the preceding measurement of the observable a on the state ψ is given by the formula

$$\mathbf{P}_\psi(b = \beta_m | a = \alpha_k) \equiv \frac{\| P^b_{\beta_m} P^a_{\alpha_k}\psi \|^2}{\| P^a_{\alpha_k} \psi \|^2} \tag{10.39}$$

One can motivate this definition by appealing to the projection postulate (Lüders' version). After the a measurement with output $a = \alpha_k$ initially prepared state ψ is projected onto the state

$$\psi_{\alpha_k} = \frac{P^a_{\alpha_k}\psi}{\| P^a_{\alpha_k} \psi \|}.$$

Then one applies Born's rule to the b measurement for this state.

10.7 Conditional Probability for Observables with a Nondegenerate Spectrum

Let the operator $\widehat{a}$ have a nondegenerate spectrum, i.e., for any eigenvalue α the corresponding eigenspace (i.e., generated by eigenvectors with $\widehat{a}\psi = \alpha\psi$) is one-dimensional. We can write

$$\mathbf{P}_\psi(b = \beta_m | a = \alpha_k) = \| P^b_{\beta_m} e^a_k \|^2 \tag{10.40}$$

(here $\widehat{a}e^a_k = \alpha_k e^a_k$).

10.7.1 Independence of the Initial State

Thus, the conditional probability in this case does not depend on the original state ψ. We can say that the memory about the original state was destroyed. If also the operator $\widehat{b}$ has a nondegenerate spectrum, then we have

$$\mathbf{P}_\psi(b = \beta_m | a = \alpha_k) = |\langle e_m^b | e_k^a \rangle|^2$$

and

$$\mathbf{P}_\psi(a = \alpha_k | b = \beta_m) = |\langle e_k^a | e_m^b \rangle|^2.$$

By using symmetry of the scalar product we obtain the following result:

10.7.2 Matrix of Transition Probabilities: Symmetric

Let both operators $\widehat{a}$ and $\widehat{b}$ have purely discrete nondegenerate spectra and let $P_k^a \psi \neq 0$ and $P_m^b \psi \neq 0$. Then conditional probability is symmetric and it does not depend on the original state ψ:

$$\mathbf{P}_\psi(b = \beta_m | a = \alpha_k) = \mathbf{P}_\psi(a = \alpha_k | b = \beta_m) = |\langle e_m^b | e_k^a \rangle|^2. \quad (10.41)$$

We remark that classical (Kolmogorov–Bayes) conditional probability is not symmetric in special situations. Thus *QM is described by a very specific probabilistic model.*

10.7.3 Matrix of Transition Probabilities: Double Stochasticity

Consider two nondegenerate observables. Set $p_{\beta|\alpha} = \mathbf{P}(b = \beta | a = \alpha)$. The matrix of transition probabilities $\mathbf{P}^{b|a} = (p_{\beta|\alpha})$ is not only *stochastic*, i.e.,

$$\sum_\beta p_{\beta|\alpha} = 1$$

but it is even *doubly stochastic*:

$$\sum_\alpha p_{\beta|\alpha} = \sum_\alpha |\langle e_\beta^b | e_\alpha^a \rangle|^2 = \langle e_\beta^b | e_\beta^b \rangle = 1.$$

In Kolmogorov's model, stochasticity is the general property of matrices of transition probabilities. However, in general classical matrices of transition probabilities are not doubly stochastic. Hence, double stochasticity is a very specific property of quantum probability.

We remark that *statistical data collected outside quantum physics, e.g., in decision making by humans and psychology, violates the quantum law of double stochasticity* [152]. Such data cannot be mathematically represented with the aid of Hermitian operators with nondegenerate spectra. One has to consider either Hermitian operators with degenerate spectra or positive-operator-valued measures (POVMs) [9].

10.8 Interference of Probabilities for Incompatible Observables

We shall show that the quantum probabilistic calculus violates the conventional FTP (10.10), one of the basic laws of classical probability theory. In this section, we proceed in an abstract setting by operating with two abstract incompatible observables. The concrete realization of this setting for the two-slit experiment demonstrating the interference of probabilities in QM will be presented in section 10.14, which is closely related to Feynman's claim [79] on the nonclassical probabilistic structure of this experiment.

Let $H_2 = \mathbf{C} \times \mathbf{C}$ be the two-dimensional complex Hilbert space and let $\psi \in H_2$ be a quantum state. Let us consider two dichotomous observables $b = \beta_1, \beta_2$ and $a = \alpha_1, \alpha_2$ represented by Hermitian operators $\widehat{b}$ and $\widehat{a}$, respectively (one may consider simply Hermitian matrices). Let $e^b = \{e^b_\beta\}$ and $e^a = \{e^a_\alpha\}$ be two orthonormal bases consisting of eigenvectors of the operators. The state ψ can be represented in the two ways:

$$\psi = c_1 e^a_1 + c_2 e^a_2, \ c_\alpha = \langle \psi | e^a_\alpha \rangle; \tag{10.42}$$

$$\psi = d_1 e^b_1 + d_2 e^b_2, \ d_\beta = \langle \psi | e^b_\beta \rangle. \tag{10.43}$$

By Postulate 4 we have

$$\mathbf{P}(a = \alpha) \equiv \mathbf{P}_\psi(a = \alpha) = |c_\alpha|^2; \tag{10.44}$$

$$\mathbf{P}(b = \beta) \equiv \mathbf{P}_\psi(b = \beta) = |d_\beta|^2. \quad (10.45)$$

The possibility of expanding one basis with respect to another basis induces a connection between the probabilities $\mathbf{P}(a = \alpha)$ and $\mathbf{P}(b = \beta)$. Let us expand the vectors e^a_α with respect to the basis e^b:

$$e^a_1 = u_{11}e^b_1 + u_{12}e^b_2 \quad (10.46)$$

$$e^a_2 = u_{21}e^b_1 + u_{22}e^b_2 \quad (10.47)$$

where $u_{\alpha\beta} = \langle e^a_\alpha, e^b_\beta \rangle$. Thus $d_1 = c_1u_{11} + c_2u_{21}$, $d_2 = c_1u_{12} + c_1u_{22}$. We obtain the *quantum rule* for transformation of probabilities:

$$\mathbf{P}(b = \beta) = |c_1u_{1\beta} + c_2u_{2\beta}|^2. \quad (10.48)$$

On the other hand, by the definition of quantum conditional probability (see (10.39)), we obtain

$$\mathbf{P}(b = \beta|a = \alpha) \equiv \mathbf{P}_\psi(b = \beta|a = \alpha) = |\langle e^a_\alpha | e^b_\beta \rangle|^2. \quad (10.49)$$

By combining Eqs. 10.44, 10.45 and 10.48, 10.49 we obtain the *quantum formula of total probability—the formula of the interference of probabilities*:

$$\mathbf{P}(b = \beta) = \sum_\alpha \mathbf{P}(a = \alpha)\mathbf{P}(b = \beta|a = \alpha)$$
$$+ 2\cos\theta\sqrt{\mathbf{P}(a = \alpha_1)\mathbf{P}(b = \beta|a = \alpha_1)\mathbf{P}(a = \alpha_2)\mathbf{P}(b = \beta|a = \alpha_2)} \quad (10.50)$$

In general, $\cos\theta \neq 0$. Thus, the quantum FTP does not coincide with the classical FTP (10.10), which is based on the Bayes' formula.

We considered only the case of observables presented by operators acting in the two-dimensional Hilbert space (see [9] for arbitrary dimensions and even representation of observables by POVMs).

10.9 Logic of Quantum Propositions

Von Neumann and Birkhoff [31, 235] suggested representing *events* (propositions) by orthogonal projectors in complex Hilbert space H.

For an orthogonal projector P, we set $H_{\mathrm{P}} = P(H)$, its image, and vice versa. For subspace L of H, the corresponding orthogonal projector is denoted by the symbol P_{L}.

The set of orthogonal projectors is a *lattice* with the order structure $P \leq Q$ if $H_{\mathrm{P}} \subset H_{\mathrm{Q}}$ or equivalently, for any $\psi \in H$, $\langle \psi | P\psi \rangle \leq \langle \psi | Q\psi \rangle$.

We recall that the lattice of projectors is endowed with operations "and" ($\wedge$) and "or" ($\vee$). For two projectors P_1, P_2, the projector $R = P_1 \wedge P_2$ is defined as the projector onto the subspace $P_{\mathrm{R}} = H_{P_1} \cap H_{P_2}$ and the projector $S = P_1 \vee P_2$ is defined as the projector onto the subspace P_{R} defined as the minimal linear subspace containing the set-theoretic union $H_{P_1} \cup H_{P_2}$ of subspaces H_{P_1}, H_{P_2}: this is the space of all linear combinations of vectors belonging these subspaces. The operation of negation is defined as the orthogonal complement $P^{\perp} = \{y \in H : \langle y|x \rangle = 0 \text{ for all } x \in H_{\mathrm{P}}\}$.

In the language of subspaces the operation "and" coincides with the usual set-theoretic intersection, but the operations "or" and "not" are nontrivial deformations of the corresponding set-theoretic operations. It is natural to expect that such deformations can induce deviations from classical Boolean logic.

Consider the following simple example. Let H be a two-dimensional Hilbert space with the orthonormal basis (e_1, e_2) and let $v = (e_1 + e_2)/\sqrt{2}$. Then, $P_v \wedge P_{e_1} = 0$ and $P_v \wedge P_{e_2} = 0$, but $P_v \wedge (P_{e_1} \vee P_{e_2}) = P_v$. Hence, for quantum events, in general the distributivity law is violated:

$$P \wedge (P_1 \vee P_2) \neq (P \wedge P_1) \vee (P \wedge P_2) \tag{10.51}$$

The lattice of orthogonal projectors is called *quantum logic.* It is considered as a (special) generalization of classical Boolean logic. Any sub-lattice consisting of commuting projectors can be treated as classical Boolean logic.

At first sight the representation of events by projectors/linear subspaces might look exotic. However, this is simply a prejudice which springs from too common usage of the set-theoretic representation of events (Boolean logic) in the modern classical probability theory. The tradition to represent events by subsets was firmly established by A. N. Kolmogorov in 1933. We remark that before him the basic classical probabilistic models were not of the set-theoretic nature. For example, the main competitor of the Kolmogorov model, the von Mises frequency model [232–234], was based on the notion of a collective.

As we have seen, quantum logic relaxes some constraints posed on the operations of classical Boolean logic, in particular the distributivity constraint. This provides novel possibilities for logically consistent reasoning; cf. with the information interpretation of QM [39–43, 48–51, 58, 59, 89–97, 138, 144–149, 175, 205–209, 238] and applications of the mathematical formalism of QM (in fact, reasoning and decision making based on quantum logic) outside of physics.

10.10 Tensor Product of Hilbert Spaces and Linear Operators

Let both state spaces be L^2 spaces, the spaces of complex-valued square integrable functions: $H_1 = L^2(\mathbb{R}^k)$ and $L^2(\mathbb{R}^m)$.

Take two functions; $\psi \equiv \psi(x)$ belongs to H_1 and $\phi \equiv \phi(y)$ belongs to H_2. By multiplying these functions we obtain the function of two variables $\Psi(x, y) = \psi(x) \times \phi(y)$, where $\times$ denotes the usual pointwise product.[4] It is easy to check that this function belongs to the space $H = L^2(\mathbb{R}^{k+m})$. Take now n functions, $\psi_1(x), \ldots, \psi_n(x)$, from H_1 and n functions, $\phi_1(y), \ldots, \phi_n(y)$, from H_2 and consider the sum of their pairwise products:

$$\Psi(x, y) = \sum_i \psi_i(x) \times \phi_i(y). \tag{10.52}$$

This function also belongs to H.

It is possible to show that any function belonging to H can be represented as Eq. 10.52, where the sum is in general infinite. Multiplication of functions is the basic example of the operation of the tensor product. The latter is denoted by the symbol $\otimes$. Thus, in the example under consideration $\psi \otimes \phi(x, y) = \psi(x) \times \phi(y)$. The tensor product structure on $H = L^2(\mathbb{R}^{k+m})$ is symbolically denoted as $H = H_1 \otimes H_2$.

---

[4] Here it is convenient to use this symbol, not just write $\Psi(x, y) = \psi(x)\phi(y)$.

Consider now two arbitrary orthonormal bases in spaces H_k, $(e_j^{(k)})$, $k = 1, 2$. Then, functions $(e_{ij} = e_i^{(1)} \otimes e_j^{(2)})$ form an orthonormal basis in H : any $\Psi \in H$, can be represented as

$$\Psi = \sum c_{ij} e_{ij} \equiv \sum c_{ij} e_i^{(1)} \otimes e_j^{(2)}, \tag{10.53}$$

where

$$\sum |c_{ij}|^2 < \infty. \tag{10.54}$$

Consider now two arbitrary finite dimensional Hilbert spaces, H_1, H_2. For each pair of vectors $\psi \in H_1$, $\phi \in H_2$, we form a new formal entity denoted by $\psi \otimes \phi$. Then we consider the sums $\Psi = \sum_i \psi_i \otimes \phi_i$. On the set of such formal sums we can introduce the linear space structure. (To be mathematically rigorous, we have to constrain this set by some algebraic relations to make the operations of addition and multiplication by complex numbers well defined). This construction gives us the tensor product $H = H_1 \otimes H_2$. In particular, if we take orthonormal bases in H_k, $(e_j^{(k)})$, $k = 1, 2$, then $(e_{ij} = e_i^{(1)} \otimes e_j^{(2)})$ form an orthonormal basis in H, any $\Psi \in H$, can be represented as (10.53) with (10.54).

The latter representation gives the simplest possibility to define the tensor product of two arbitrary (i.e., may be infinite-dimensional) Hilbert spaces as the space of formal series (10.53) satisfying the condition (10.54).

Besides the notion of the tensor product of states, we shall also use the notion of the tensor product of operators. Consider two linear operators $\widehat{a}_i : H_i \to H_i$, $i = 1, 2$. Their tensor product $\widehat{a} \equiv \widehat{a}_1 \otimes \widehat{a}_2 : H \to H$ is defined starting with the tensor products of two vectors:

$$\widehat{a}\ \psi \otimes \phi = (\widehat{a}_1 \psi) \otimes (\widehat{a}_2 \phi).$$

Then it is extended by linearity. By using the coordinate representation (10.53) the tensor product of operators can be represented as

$$\widehat{a}\ \Psi = \sum c_{ij}\ \widehat{a} e_{ij} \equiv \sum c_{ij} \widehat{a}_1 e_i^{(1)} \otimes \widehat{a}_2 e_j^{(2)}, \tag{10.55}$$

If operators $\widehat{a}_i$, $i = 1, 2$, are represented by matrices $(a_{kl}^{(i)})$, with respect to the fixed bases, then the matrix $(a_{kl.nm})$ of the operator

$\widehat{a}$ with respect to the tensor product of these bases can be easily calculated.

In the same way, one defines the tensor product of Hilbert spaces $H_1, \ldots, H_n$ denoted by the symbol $H = H_1 \otimes \ldots \otimes H_n$. We start with forming the formal entities $\psi_1 \otimes \ldots \otimes \psi_n$, where $\psi_j \in H_j$, $j = 1, \ldots, n$. Tensor product space is defined as the set of all sums $\sum_j \psi_{1j} \otimes \ldots \otimes \psi_{nj}$ (which has to be constrained by some algebraic relations, but we omit such details). Take orthonormal bases in H_k, $(e_j^{(k)})$, $k = 1, \ldots, n$. Then, any $\Psi \in H$ can be represented as

$$\Psi = \sum_{\alpha} c_\alpha e_\alpha \equiv \sum_{\alpha=(j_1 \ldots j_n)} c_{j_1 \ldots j_n} e_{j_1}^{(1)} \otimes \ldots \otimes e_{j_n}^{(n)}, \qquad (10.56)$$

where $\sum_\alpha |c_\alpha|^2 < \infty$.

10.11 Ket and Bra Vectors: Dirac's Symbolism

Dirac's notations [65] are widely used in quantum information theory. Vectors of H are called *ket vectors* and are denoted as $|\psi\rangle$. The elements of the dual space H' of H, the space of linear continuous functionals on H, are called *bra vectors* and are denoted as $\langle\psi|$.

Originally, the expression $\langle\psi|\phi\rangle$ was used by Dirac for the duality form between H' and H, i.e., $\langle\psi|\phi\rangle$ is the result of application of the linear functional $\langle\psi|$ to the vector $|\phi\rangle$. In mathematical notation it can be written as follows. Denote the functional $\langle\psi|$ by f and the vector $|\phi\rangle$ by simply ϕ. Then, $\langle\psi|\phi\rangle \equiv f(\phi)$. To simplify the model, Dirac later took the assumption that H is a Hilbert space, i.e., H' can be identified with H. We remark that this assumption is an axiom simplifying the mathematical model of QM. However, in principle, Dirac's formalism [65] is applicable for any topological linear space H and its dual space H', so it is more general than von Neumann's formalism [235] rigidly based on Hilbert space.

Consider an observable a given by the Hermitian operator $\widehat{a}$ with a nondegenerate spectrum and restrict our consideration to the case of finite-dimensional H. Thus, the normalized eigenvectors e_i of A form the orthonormal basis in H. Let $\widehat{a}e_i = \alpha_i e_i$. In Dirac's notation

e_i is written as $|\alpha_i\rangle$ and, hence, any pure state can be written as

$$|\psi\rangle = \sum_i c_i|\alpha_i\rangle, \; \sum_i |c_i|^2 = 1. \tag{10.57}$$

Since the projector onto $|\alpha_i\rangle$ is denoted as $P_{\alpha_i} = |\alpha_i\rangle\langle\alpha_i|$, the operator $\widehat{a}$ can be written as

$$\widehat{a} = \sum_i \alpha_i|\alpha_i\rangle\langle\alpha_i|. \tag{10.58}$$

Now consider two Hilbert spaces H_1 and H_2 and their tensor product $H = H_1 \otimes H_2$. Let $(|\alpha_i\rangle)$ and $(|\beta_i\rangle)$ be orthonormal bases in H_1 and H_2 corresponding to the eigenvalues of two observables A and B. Then, vectors $|\alpha_i\rangle \otimes |\beta_j\rangle$ form the orthonormal basis in H. Typically in physics the sign of the tensor product is omitted and these vectors are written as $|\alpha_i\rangle|\beta_j\rangle$ or even as $|\alpha_i\beta_j\rangle$. Thus, any vector $\psi \in H = H_1 \otimes H_2$ can be represented as

$$\psi = \sum_{ij} c_{ij}|\alpha_i\beta_j\rangle, \tag{10.59}$$

where $c_{ij} \in \mathbf{C}$ (in the infinite-dimensional case these coefficients are constrained by the condition $\sum_{ij} |c_{ij}|^2 < \infty$).

10.12 Quantum Bit: Using State Superposition for Information Encoding

In particular, in quantum information theory, typically qubit states are represented with the aid of observables having the eigenvalues 0, 1. Each qubit space is two-dimensional:

$$|\psi\rangle = c_0|0\rangle + c_1|1\rangle, \; |c_0|^2 + |c_1|^2 = 1. \tag{10.60}$$

A pair of qubits is represented in the tensor product of single qubit spaces. Here, pure states can be represented as superpositions:

$$|\psi\rangle = c_{00}|00\rangle + c_{01}|01\rangle + c_{10}|10\rangle + c_{11}|00\rangle, \tag{10.61}$$

where $\sum_{ij} |c_{ij}|^2 = 1$. In the same way the n-qubit state is represented in the tensor product of n one-qubit state spaces (it has the dimension 2^n):

$$|\psi\rangle = \sum_{x_j=0,1} c_{x_1 \ldots x_n}|x_1 \ldots x_n\rangle, \tag{10.62}$$

where $\sum_{x_j=0,1} |c_{x_1 \ldots x_n}|^2 = 1$. We remark that the dimension of the n-qubit state space grows exponentially with the growth of n. The natural question about possible physical realizations of such multidimensional state spaces arises. The answer to it is not completely clear; it depends very much on the used interpretation of the wave function.

10.13 Entanglement of Pure and Mixed Quantum States

Consider the tensor product $H = H_1 \otimes H_2 \otimes \ldots \otimes H_n$ of Hilbert spaces H_k, $k = 1, 2, \ldots, n$. The states of the space H can be *separable and nonseparable* (entangled). We start by considering pure states. The states from the first class, separable pure states, can be represented in the form

$$|\psi\rangle = \otimes_{k=1}^{n} |\psi_k\rangle = |\psi_1 \ldots \psi_n\rangle, \qquad (10.63)$$

where $|\psi_k\rangle \in H_k$. The states that cannot be represented in this way are called nonseparable, or *entangled.* Thus, mathematically, the notion of entanglement is very simple: it means impossibility of tensor factorization.

For example, let us consider the tensor product of two one-qubit spaces. Select in each of them an orthonormal basis denoted as $|0\rangle, |1\rangle$. The corresponding orthonormal basis in the tensor product has the form $|00\rangle, |01\rangle, |10\rangle, |11\rangle$. Here, we used Dirac's notations (see section 10.11) near the end. Then, so-called Bell's states

$$|\Phi^+\rangle = (|00\rangle + |11\rangle)/\sqrt{2}, \quad |\Phi^-\rangle = (|00\rangle - |11\rangle)/\sqrt{2}; \qquad (10.64)$$

$$|\Psi^+\rangle = (|01\rangle + |10\rangle)/\sqrt{2}, \quad |\Psi^-\rangle = (|01\rangle - |10\rangle)/\sqrt{2}, \qquad (10.65)$$

are entangled.

Although the notion of entanglement is mathematically simple, its physical interpretation is one of the main problems of modern quantum foundations. The common interpretation is that entanglement encodes quantum nonlocality, the possibility of action at a distance (between parts of a system in an entangled state). Such an interpretation implies drastic change in all classical physical presentations about nature, at least about the microworld. In probabilistic terms, entanglement induces correlations which are too strong to be described by classical probability theory. (At least this is the common opinion of experts in quantum information theory and quantum foundations.) Such correlations violate the famous Bell inequality, which can be derived only in a classical probability framework. The latter is based on the use of a single probability space covering probabilistic data collected in a few incompatible measurement contexts.

Now consider a quantum state given by density operator ρ in H. This state is called separable if it can be factorized in the product of density operators in spaces H_k:

$$\rho = \otimes_{k=1}^{n} \rho_k. \tag{10.66}$$

Otherwise the state ρ is called entangled. We remark that an interpretation of entanglement for mixed states is even more complicated than for pure states.

10.14 Two-Slit Experiment and Violation of the Classical Law of Total Probability

10.14.1 On the Possibility of Classical Probabilistic Description of Quantum Experiments

In this section we discuss the old foundational problem of QM: *whether it is possible to represent quantum states by classical probability (CP) distributions and quantum observables by random variables* [197]. Since the first days of QM, it was commonly believed that classical probability theory (based on Kolmogorov's axiomatics [191]) cannot serve to represent incompatible quantum observables. At the early stage of development of QM this belief was firmly based on *the Heisenberg uncertainty principle.* By a straightforward interpretation of the Heisenberg uncertainty relation, position and momentum cannot be jointly assigned to an individual quantum system. Under such an interpretation, it was meaningless to even speak about the joint probability distribution (jpd) for position and momentum. For example, for Wigner [237], it was clear that

> In quantum theory there does not exist any similar simple expression for the probability, because one cannot ask for the simultaneous probability for the coordinates and momenta. (p. 749)

Here, "similar" is related to the case of classical statistical mechanics and the Gibbs–Boltzmann formula for statistical equilibrium.

The *two-slit* experiment is the basic example demonstrating that QM describes statistical properties in microscopic phenomena, to which the classical probability theory *seems to be not applicable* (see Feynman [78]) (italic shrift was added by the authors of this paper):

> From about the beginning of the twentieth century experimental physics amassed an impressive array of strange phenomena which demonstrated the inadequacy of classical physics. The attempts to discover a theoretical structure for the new phenomena led at first to a confusion in which it appeared that light,and electrons, sometimes behaved like waves and sometimes like particles. This apparent inconsistency was completely resolved in 1926 and 1927 in the theory called quantum mechanics. The new theory asserts that there are experiments for which the exact outcome is fundamentally unpredictable, and that in these cases one has to be satisfied with computing probabilities of various outcomes. *But far more fundamental was the discovery that in nature the laws of combining probabilities were not those of the classical probability theory of Laplace.*

The latter statement of Feynman is based on his analysis of the probabilistic structure of the two-slit experiment [78, 79]. This analysis demonstrated (at least as Feynman believed) that probabilities associated with this experiment violate the *law of additivity of probability*, the basic law of classical probability theory. In a series of authors' works [133, 136, 141, 145, 153, 154], Feynman's analysis [79] was embedded into the contextual framework, i.e., treatment of the two-slit experiment as a bunch of experiments related to different experimental contexts (see Figs. 10.2, 10.3, 10.4). This contextual probabilistic analysis led to the conclusion that the probabilities for the two-slit experiment (in fact, a bunch of experiments) collected experimentally and predicted by QM violate FTP (section 10.1, Eq. 10.10). In the abstract framework for incompatible observables, this statement was proven in section 10.8. QM predicts a modified formula of total probability, FTP with the *interference term.*

We remark that Feynman's no-go conclusion [79] based on consideration of the two-slit experiment is justified better than the original claims based on the Heisenberg uncertainty relation. The role of Heisenberg's uncertainty relation for quantum foundations is a complicated topic. On the one hand, this relation played a crucial role in establishing the complementarity principle by Bohr. On the other, it is in general interpreted too straightforwardly as carrying the message about the impossibility of the joint determination of the position and momentum for an *individual quantum system.* As was rightly pointed by Margenau and Ballentine [17–19] (see, e.g., [153] for discussion, Heisenberg's uncertainty relation involves the standard deviations for the position and momentum observables. These deviations are measured in two separate experiments, for the position and momentum observables, respectively. It seems that such separate measurements have nothing to do with joint measurability of these observables. Therefore, this relation cannot be used to justify the claim that the classical probability distribution for these observables does not exist.

In any event the problem of the interrelation of classical and quantum probabilities is very complex. Of course, as was already remarked, for one concrete quantum observable, the statistics of its outcomes is well described by classical probability theory. The main problem is whether it is possible to use classical probability to describe the statistics of outcomes of incompatible quantum observables. On the one hand, the contextual structure of measurements and the impossibility to combine the contexts of incompatible measurements lead to the *contextual probabilistic description* [133, 136, 141, 148, 151, 153, 172], which is not covered straightforwardly by the Kolmogorov probability model. For example, as was shown in section 10.1, classical FTP is violated in QM formalism. In section 10.14.2, we shall illustrate this situation by the two-slit experiment. On the other hand, the contextual decomposition of probability is not totally incompatible with Kolmogorov's theory. In principle, incompatible contexts can be embedded into the common Kolmogorov probability space, but not straightforwardly.

The first classical probabilistic representation of QM based on the *symplectic tomogram* was constructed [66, 195, 196]. Another construction of the classical probabilistic representation of QM is based on the so-called *prequantum classical statistical field theory* [157–160, 162, 174]. Representation of the Bohm–Bell-type experiments based on classical conditional probabilities was constructed [13, 165, 179]; see [164] for application of such a technique to the two-slit experiment. Recently, Dzhafarov et al. [70–72] (and especially article [73] and the comment on the latter by Khrennikov [183]) proposed a classical probabilistic description of the basic QM experiments based on the *coupling technique* which widely used in classical probability.

10.14.2 Interference of Wave Functions

In this section, we consider the experiment with the symmetric setting: the source of photons is located symmetrically with respect to two slits (Fig. 10.1).

Consider the pair of observables a and b. We select a as the "slit passing observable," i.e., $a = 0, 1$ (Fig. 10.1) (we use indexes 0, 1 to be close to qubit notation) and observable b as the position on the photosensitive plate (Fig. 10.2). We remark that the b-observable has a continuous range of values, the position x on the photosensitive plate. We denote $\mathbf{P}(a = i)$ by $\mathbf{P}(i)$ $(i = 0, 1)$, and $\mathbf{P}(b = x)$ by $\mathbf{P}(x)$. Physically, the a-observable corresponds to measurement of position (coarse-grained to "which slit?") and the b-observable represents the measurement of momentum.

In quantum foundational studies, various versions of the two-slit experiment have been successfully performed not only with photons but also with electrons and even with macroscopic molecules (by

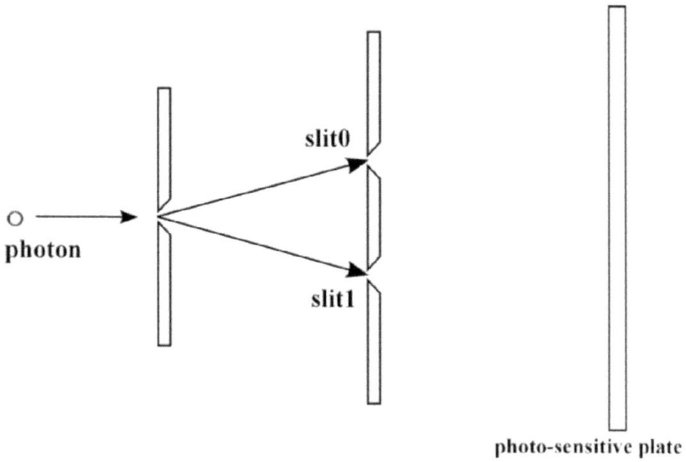

Figure 10.1 Experimental setup.

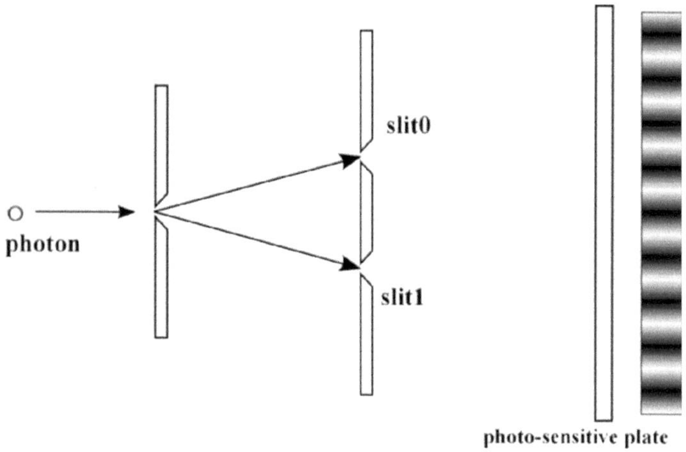

Figure 10.2 Context with both slits open.

Zeilinger's group). All those experiment demonstrated matching with predictions of QM. Experimenters reproduce the interference patterns predicted by QM and calculated by using wave functions.

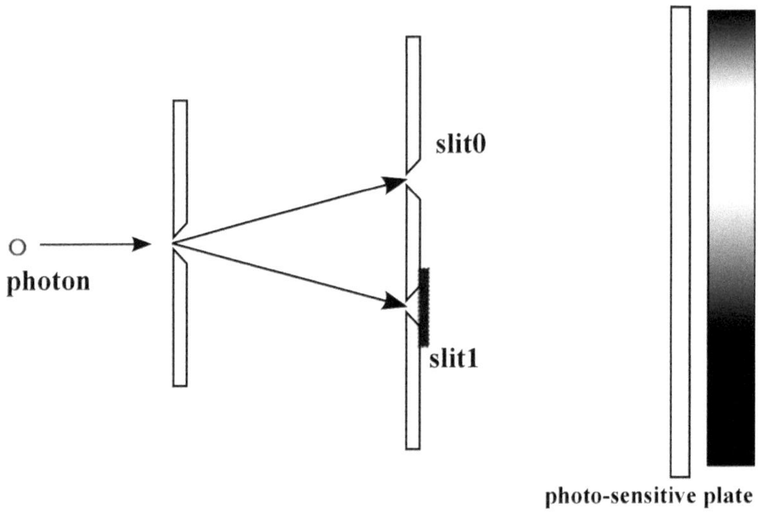

Figure 10.3 Context with one slit closed-I.

The probability that a photon is detected at position x on the photosensitive plate is represented as

$$\begin{aligned}\mathbf{P}(x) &= \left|\frac{1}{\sqrt{2}}\psi_0(x) + \frac{1}{\sqrt{2}}\psi_1(x)\right|^2 \\ &= \frac{1}{2}\left|\psi_0(x)\right|^2 + \frac{1}{2}\left|\psi_1(x)\right|^2 + \left|\psi_0(x)\right|\left|\psi_1(x)\right|\cos\theta, \qquad (10.67)\end{aligned}$$

where ψ_0 and ψ_1 are two wave functions, whose squared absolute values $|\psi_i(x)|^2$ give the distributions of photons passing through the slit $i = 0, 1$, see Figs. 10.3 and 10.4. Here, we explored the rule of addition of complex probability amplitudes, a quantum analogue of the rule of addition of probabilities. This rule is the direct consequence of the linear space structure of quantum state spaces.

The term

$$|\psi_0(x)|\,|\psi_1(x)|\cos\theta$$

implies the interference effect of two wave functions. Let us denote $|\psi_i(x)|^2$ by $\mathbf{P}(x|i)$. Then, Eq. 10.67 is represented as

$$\begin{aligned}\mathbf{P}(x) = \mathbf{P}(0)\mathbf{P}(x|0) + \mathbf{P}(1)\mathbf{P}(x|1) \qquad (10.68) \\ + 2\sqrt{\mathbf{P}(0)\mathbf{P}(x|0)\mathbf{P}(1)\mathbf{P}(x|1)}\cos\theta.\end{aligned}$$

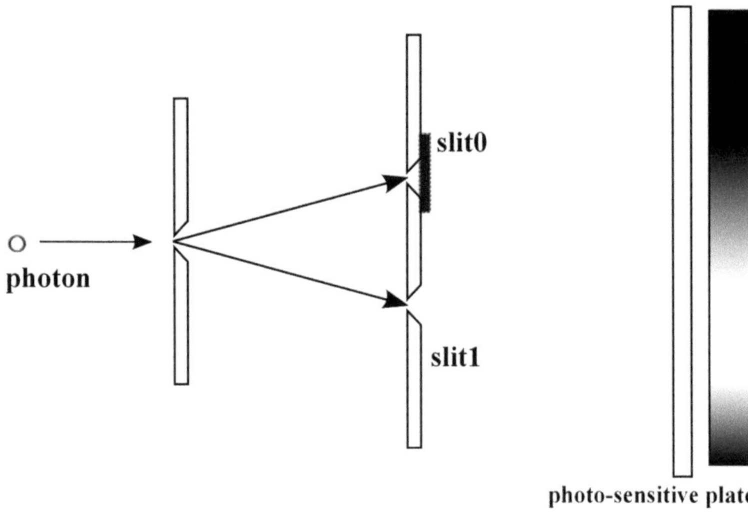

Figure 10.4 Context with one slit is closed-II.

Here, the values of probabilities $\mathbf{P}(0)$ and $\mathbf{P}(1)$ are equal to 1/2 since we consider the symmetric settings. For general experimental settings, $\mathbf{P}(0)$ and $\mathbf{P}(1)$ can be taken as the arbitrary nonnegative values satisfying $\mathbf{P}(0) + \mathbf{P}(1) = 1$. In the above form, the classical probability law (FTP)

$$\mathbf{P}(x) = \mathbf{P}(0)\mathbf{P}(x|0) + \mathbf{P}(1)\mathbf{P}(x|1), \tag{10.69}$$

is violated, and the term of interference

$$2\sqrt{\mathbf{P}(x|0)\mathbf{P}(0)\mathbf{P}(x|1)\mathbf{P}(1)}\cos\theta$$

specifies the violation.

The crucial point is that the two-slit experiment has a multicontextual structure with three different contexts:

(1) C_0: only the $i = 0$th slit is open (Fig. 10.3)
(2) C_1: only the $i = 1$th slit is open (Fig. 10.4)
(3) C_{01}: both slits are open (Fig. 10.2)

Comparison of possibilities is represented as comparison of the corresponding probability distributions $\mathbf{P}(x|i)$, $\mathbf{P}(x)$. In the contextual notations they can be written as

$$p^b_{C_i}(x) \equiv \mathbf{P}(b = x|C_i),\ p^b_{C_{01}}(x) \equiv \mathbf{P}(b = x|C_{01}).$$

Here, conditioning is not classical probabilistic event conditioning, but context conditioning [153]: different contexts are mathematically represented by different Kolmogorov probability spaces. The general contextual probability theory including its representation in a complex Hilbert space is presented in detail in my monograph [153].

Chapter 11

QBism: Subjective Probabilistic Interpretation of Quantum Mechanics

This chapter is devoted to quantum Bayesianism (QBism), the subjective probability interpretation of QM. Creation of such an interpretation is an important step towards unification of applications of the mathematical formalism of QM to physical and to cognitive and sociopolitical systems. QBism can serve to couple the general theory of decision making based on subjective probability with quantum formalism. On the other hand, our quantum-like modeling of sociopolitical processes does not depend on interpretations of QM. Therefore, the reader need not read this chapter, which has a merely foundational value.

As was stressed in section 10.5, QM, which manifests huge success in the mathematical representation of the basic problems of physics of the microworld, suffers heavily from diversity of interpretations of this mathematical representation. QBism is one of the most recent attempts to provide a consistent interpretation of QM, free of all possible mysteries and paradoxes. In this chapter *I present my personal reflections on QBism* (see also [177]). The discussion that follows is definitely not systematic enough to serve as an introduction to QBism. Readers who are already conversant

Social Laser
Andrei Khrennikov

ISBN 978-981-4800-83-9 (Hardcover), 978-0-367-81720-6 (eBook)
www.jennystanford.com

with QBism are unlikely to find anything new in our presentation. These notes have two closely related aims:

(1) to add some details to the history of the foundation and evolution of QBism
(2) to send a message to that part of the physical community which still strongly rejects QBism as a totally unphysical interpretation of QM

It is important to point out that the present chapter is written by one of the active representatives of this anti-QBist lobby which still strongly dominates in quantum foundations. I definitely have not been completely converted into QBism, but recently I started to find rational points in QBists? views on quantum theory, and I want to present these points, but, so to say, with a sense of doubt. This chapter starts with a historical remark on the first years of QBism. It continues with a brief representation of its essentials, mainly referring to and citing its creators. Then the main postulates of QBism are critically analyzed.

We point out that the cornerstone of QBism is the interpretation of QM as a machine for updating probabilities based on a modification of FTP (the classical formula of total probability [see section 10.1], equality (10.10)). We emphasize this mathematically probabilistic dimension of QBism which is typically shadowed by the emphasis on the subjective interpretation of probability and private agent perspective to quantum measurements. This emphasis often creates the impression that QBism is merely about a philosophy of quantum mechanics. At least this is my personal experience based on hundreds of conversations about QBism with leading experts in quantum foundations. It is important to stress that QBism's version of FTP is very special and it is derived with the aid of *symmetric informationally complete positive-operator-valued measures* (SIC-POVMs) (see section 11.3).

From this viewpoint, i.e., QM as a probability updating machine, QBism is similar to the Växjö interpretation of QM. The latter is a realist contextual statistical interpretation, so ideologically it is opposite to QBism. However, the probability update basis makes it close to QBism. At the same time, even this closeness is only formal, since the QBism version of generalized FTP differs crucially from the Växjö version [133, 136, 141, 145, 153, 154].

This choice of the quantum modification of FTP is critically analyzed. The main output of this analysis is that it seems that QBists are really addicted to exploration of SIC-POVMs (see section 11.3). And it seems that QBism would earn a lot by proceeding without the assumption of the symmetry of IC-POVM or maybe even without appealing to information completeness.

In this chapter QBism is compared with two interpretations of QM which are very close to QBism, but in two totally different aspects; cf. [87]. In section 11.4 we compare QBism with the Växjö interpretation [137, 144, 153] of QM. The latter is a contextual realistic interpretation based on the objective interpretation of probability (with sympathy [131, 152] to the von Mises frequency probability theory [232–234]). At first sight, there is nothing in common between these two interpretations (see [90] and [137] for a critical debate). However, they have one very important thing in common: in both interpretations the Born rule is treated as a quantum modification of classical FTP, although the modifications of FTP are mathematically very different. In section 11.7 QBism is compared with the information interpretation of QM elaborated by A. Zeilinger and C. Brukner ([39–42, 238]). At first sight, these interpretations are very similar. However, we shall show that, in fact, they differ crucially.

The rest of this chapter is devoted to interpretations of probability. Here diversity is not so huge as in QM, but the gap between two main interpretations, objective and subjective, is no smaller than the gap between two basic trends in interpretations of QM, realist (in the spirit of Einstein) and non-realist (in the spirit of Copenhagen). We present the original Kolmogorov interpretation of probability [191] (which is not so well known, even in the probability community) and compare it with the genuine frequency interpretation of von Mises [232–234] and subjective interpretation of de Finetti [62, 63].

Then, following de Finetti, we point out that consistent appealing to the subjective interpretation of probability should lead to reconsideration of the objective treatment of the scientific methodology. Finally, we point that de Finetti was even more revolutionary than

QBists, because his subjective treatment of scientific method was not restricted to a "special quantum world." In some way, QBists made one good turn but refrained from another. They revolutionarily declared the private agent (user) perspective to knowledge about the quantum world, but were not brave enough to follow de Finetty completely, i.e., to declare the private agent perspective for knowledge about the classical world as well. The latter is an exciting project still awaiting its realization.

As was pointed out at the very beginning of this chapter, it presents *critical reflections about QBism.* In particular, some doubt is expressed regarding how much the QBists' subjectivism is actually important for the overall project. I note that there may be some sort of rapprochement between subjectivism and objectivism. Of course, this may significantly underplay the role of the QBists' subjectivity in the resolution of the old puzzles about measurement and locality in quantum mechanics. At the same time, it may stimulate QBists to concentrate their efforts on development of the "probability updating machine" dimension of QBism. The same can be said about the critique of another basic element of QBism: derivation of SIC-POVM (see section 11.3) based modification of FTP.

11.1 QBism in Växjö

In 2001 QBism was strongly represented at the second Växjö conference on quantum foundations, "Quantum Theory: Reconsideration of Foundations" (QTFT2001), June 17–21, 2001. We (organizers and participants of this conference) strongly believed that the quantum information revolution would soon lead to a great foundational revolution. Unfortunately, the dream did not come true. Nevertheless, the energy of the quantum information revolution was transformed in a series of stormy debates during a series of the Växjö conferences, 2000–2019. Although these debates did not lead to a complete resolution of the basic problems of quantum mechanics, they clarified some of these problems, especially the problem of the interpretation of a quantum state. QBism was definitely one of the main foundational outputs of the quantum information revolution.

Besides QBism, we can mention the *Växjö interpretation* of QM (statistical realist and contextual) [137, 144] derivation of QM formalism from simple operational principles, D' Ariano [58, 59] and Chiribella et al. [50, 51] (for the first time this project was also announced in Växjö), and the *statistical Copenhagen interpretation* (statistical nonrealist), the final formulation of which was presented at the Växjö-15 conference by A. Plotnitsky and based on his previous studies about the probabilistic structure of QM [205–208].

As the organizers of QTFT 2001, both C. Fuchs and I dreamed for the creation of a consistent and clear interpretation of QM, free of mysteries and paradoxes. However, we went in two opposite directions. I followed Einstein and later, as a result of better understanding of Bohr's writings (and especially the comments of A. Plotnitsky on them), tried to unify Einstein's realist statistical interpretation with Bohr's contextual interpretation by filtering out Bohr's nonrealist attitude (see [137, 144, 153] for the so-called *Växjö interpretation* of QM. Both Einstein and Bohr (as well as, e.g., von Neumann) used the statistical interpretation of quantum probabilities. Therefore, the Växjö interpretation is based on this statistical interpretation. It also was important that as student I was strongly influenced by A. N. Kolmogorov and B. V. Gnedenko, who always emphasized that probability is objective and statistical. Later, after my PhD, I discovered the works of von Mises [232–234]. I really enjoyed this reading! Von Mises's interpretation of probability differs from Kolmogorov's interpretation. Nevertheless, Misesian probability is also objective and statistical. In any event, none of them (Einstein, Bohr, von Neumann, Kolmogorov, Gnedenko), including me, would agree with De Finetti's slogan:

Probability does not exist!

and with his subjective interpretation of probability. My views on the interpretation of quantum states and probabilities were presented in [137].

C. Fuchs went in the opposite direction: he (with the support of C. M. Caves, R. Schack, and D. Mermin) openly, loudly, and proudly declared [48, 49, 91–97] that QM is only about knowledge (and here QBists are very close to the fathers of the

Copenhagen interpretation, N. Bohr and W. Heisenberg). But this widely supported viewpoint was completed with very strained and revolutionary declaration that this quantum knowledge has to be treated as *personal knowledge*. Subjective interpretation of quantum probability perfectly matches such a private agent perspective of quantum theory. QBists emphasize the Bayesian probability update and decision making structure of the quantum probability calculus.

From my viewpoint, the latter is one of the main contributions of QBism in the clarification of quantum foundations. Independently this viewpoint was presented in the framework of the Växjö interpretation (section 11.4) [137, 144, 153].

At the beginning, C. Fuchs and I did not recognize this similarity, namely, the interpretation of the calculus of quantum probabilities as a machinery for updating probabilities. And it is clear why: at that time I emphasized realism and objectivity, and C. Fuchs privacy and subjectivity. And the probability update dimension was shadowed by these philosophical issues. This explains the appearance of Fuch's anti-Växjö paper [90].

Nevertheless, intuitively I felt sympathy to QBism, but the roots of this sympathy were not clear for me.[1] At the Växjö 15 conference QBism was widely represented and celebrated for its worldwide recognition. Nobel Prize Laureate T. Hänsch presented a great lecture about QBism as the only possible consistent foundational basis of quantum information theory. This lecture ignited a stormy debate, and T. Hänsch and C. Fuchs were attacked by the realist opposition (led by L. Vaidman and A. Elitzur).

11.2 Quantum Theory as Subjective Probability Machinery

In contrast to Bohr and von Neumann, Fuchs proposed to interpret probability in a subjective way. To present the essentials of some

[1] Once Christopher Fuchs asked me: "Why did you support QBism so strongly during the Växjö series of conferences? QBism contradicts your own Växjö interpretation!" In fact, I was not able to explain this even to myself. I had a feeling that QBIsm can be useful. But how? and where?

theory, sometimes it is practical to simply cite the works of its creators (this is definitely not the case with Bohr, or von Neumann, or even myself). Here we cite C. Fuchs and R. Schack [96] (pp. 3, 4):

> The fundamental primitive of QBism is the concept of experience. According to QBism, quantum mechanics is a theory that any agent can use to evaluate her expectations for the content of her personal experience.
>
> QBism adopts the personalist Bayesian probability theory pioneered by Ramsey [214] and de Finetti [63] and put in modern form by Savage [216] and Bernardo and Smith [26] among others. This means that QBism interprets all probabilities, in particular those that occur in quantum mechanics, as an agent's personal, subjective degrees of belief. This includes the case of certainty—even probabilities 0 or 1 are degrees of belief. . . .
>
> In QBism, a measurement is an action an agent takes to elicit an experience. The measurement outcome is the experience so elicited. The measurement outcome is thus personal to the agent who takes the measurement action. In this sense, quantum mechanics, like probability theory, is a single user theory. A measurement does not reveal a pre-existing value. Rather, the measurement outcome is created in the measurement action.
>
> According to QBism, quantum mechanics can be applied to any physical system. QBism treats all physical systems in the same way, including atoms, beam splitters, Stern–Gerlach magnets, preparation devices, measurement apparatuses, all the way to living beings and other agents. In this, QBism differs crucially from various versions of the Copenhagen interpretation. . . .
>
> An agent's beliefs and experiences are necessarily local to that agent. This implies that the question of nonlocality simply does not arise in QBism.

We shall revisit the interpretational issues of QBism after the presentation of its basic probabilistic principle in the next section.

The previous citations might create the impression that the subjective interpretation of quantum probabilities is the key point of QBism. It might be that even its creators have the same picture of their theory. For me, the essence of QBism is neither this special interpretation of quantum probabilities nor the concrete agent perspective of QM. For me, the main ideological invention of C. Fuchs and R. Schack was a treatment of the mathematical formalism of QM as a generalization of the classical Bayesian machinery of the probability update. This viewpoint clarifies the meaning of the basic rule of QM: the Born rule as a complex Hilbert space representation of the generalization of FTP, the *classical formula of total probability*.

This dimension of QBism is identical to the probability update dimension of the Växjö interpretation [137, 144, 153]. There are a few differences.

One is the difference in the interpretations of probability. However, nowadays I do not consider it as a crucial difference (in contrast to my first debates with C. Fuchs in 2001–2003 [90, 137]). Really, in a long series of updates the subjective and statistical viewpoints coincide.

In the debates mentioned, I too was strongly against the antirealist attitude of QBism. However, now this attitude does not disturb me as much as it did 12 years ago. Either I started to understand the QBists? views on the problem of realism better or the QBists changed their views (or both). QBism need not appeal to any subquantum model providing an ontic description of quantum systems and processes, in particular, to hidden variables. Nor is QBism concerned with struggle against such models. It seems that the personal position of C. Fuchs is similar to that of N. Bohr: for quantum *physics,* it plays no role whether finally one would be able to construct a realistic subquantum model or not. For the present state of development of quantum theory, this is the most reasonable position, compared with Zeilinger's strong antirealist attitude. Moreover, just recently (through a series of email exchanges) I understood better the position of C. Fuchs on the problem of non/realism. Surprisingly, QBists (at least C. Fuchs) do not consider QBism as a nonrealist interpretation of QM.

For me, the main difference between QBism and the Växjö interpretation is in the mathematics, not in physics or in philosophy. Both the Växjö interpretation of QM [134, 135, 153] and QBism [92–97] *treat the Born rule as a generalization of the classical formula of total probability* (FTP) in the language of linear operators.

However, these interpretations are based on two totally different mathematical generalizations of FTP, both matching the Born rule.

Starting with the Born rule, QBists derived their special version of generalized FTP which is based on a special class of the quantum probability updates, based on atomic instruments with SIC-POVMs (see section 11.3).

We now briefly present the QBism scheme for the probability update, namely the representation of the Born rule as a generalization of FTP; here we again follow Fuchs and Schack [92–96]. We remark that in coming considerations the interpretation of probabilities does not play any role. They can be subjective probabilities (as originally in QBism), but they also can be statistical, e.g., Kolmogorovian or Misesian, as well.

Quantum states are represented by density operators ρ in a Hilbert space assumed to be finite-dimensional. A measurement (an action taken by the agent) is described by a POVM $F = (F_j)$, where j labels the potential outcomes experienced by the agent. The agent's personalist probability[2] $p(F_j)$ of experiencing outcome j is given by the Born rule:

$$p(F_j) = TrF_j\rho. \tag{11.1}$$

Similar to the probabilities on the left-hand side of the Born rule, QBism regards the operators ρ and F_j on the right-hand side as judgments made by the agent, representing her personalist degrees of belief.

The Born rule as written in Eq. 11.1 appears to connect probabilities on the left-hand side of the equation with other kinds of mathematical objects—operators—on the right-hand side.

QBists assume that the agent's reference measurement is an arbitrary, *informationally complete POVM*, $E = (E_i)$ (see section 11.3 for the definition), such that each E_i is of rank 1, i.e., is proportional to a one-dimensional projector. Such measurements exist for any finite Hilbert-space dimension. Furthermore, we assume that, if the agent carries out the measurement $E = (E_i)$ for an initial state ρ,

[2]As was remarked, probability can be interpreted in other ways. The situation is similar to the classical probability update. De Finetti would treat this probability as subjective, but Kolmogorov, or Gnedenko, or von Mises, as statistical.

upon getting outcome E_i he would update to the postmeasurement state

$$\rho_i = \frac{E_i \rho E_i}{Tr E_i \rho E_i}.$$

This is the assumption of atomicity of this quantum instrument. This is a strong constraint on the class of instruments which are used in QBism. It would be interesting to analyze the possibility to proceed with arbitrary instruments.

Because the reference measurement is informationally complete, any state ρ corresponds to a unique vector of probabilities

$$p(E_i) = Tr E_i \rho,$$

and any POVM $F = (F_j)$ corresponds to a unique matrix of conditional probabilities

$$p(F_j|E_i) = tr(F_j \rho_i).$$

The operators ρ and F_j on the right-hand side of the Born rule are thus mathematically equivalent to sets of probabilities $p(E_i)$ and conditional probabilities $p(F_j|E_i)$. (We remark that all these probabilities depend on the state ρ, i.e., $p(E_i) \equiv p_\rho(E_i)$, $p(F_j) \equiv p_\rho(F_j)$, $p(F_j|E_i) = p_\rho(F_j|E_i)$).

Then, Fuchs and Schack [92, 96] stress that POVMs as well as quantum states represent the agent's personal degrees of belief. However, this is not essential for a formal scheme of probability update. We can as well interpret the probabilities $p(E_i)$ and $p(F_j|E_i)$ statistically. The main point which was rightly emphasized by them is that the Born rule can be interpreted as one special form of transformation of probabilities:

$$p(F_j) = f(p(E_i), p(F_j|E_i)). \tag{11.2}$$

Comparing with the classical FTP which is, in these notations, written as

$$p(F_j) = \sum_i p(E_i) p(F_j|E_i) \tag{11.3}$$

they formulate the statement that I consider as the cornerstone of QBism:

The Born rule is one of the forms of generalization of FTP.

For me, the main problem of QBists is that they started with a SIC-POVM $E = (E_i)$. They say [92]:

> *The Born rule allows the agent to calculate her outcome probabilities $p(F_j)$ in terms of her probabilities $p(E_i)$ and $p(F_j|E_i)$ defined with respect to a counterfactual reference measurement.*

This reference to counterfactuals is really redundant. Why should SIC-POVM measurement appear at all? As in the Växjö approach, one can start with an arbitrary POVM measurement, say $G = (G_i)$, to define probabilities $p(G_i) = Tr\rho G_i$ providing information about the state ρ.[3] By taking into account such a possibility of generalization of QBist consideration[4] I completely agree with the following statement of QBism [92]:

> *In QBism, the Born rule functions as a coherence requirement. Rather than setting the probabilities $p(F_j)$, the Born rules relates them to those defining the state ρ and the POVM $F = (F_j)$. Just like the standard rules of probability theory, the Born rule is normative: the agent ought to assign probabilities that satisfy the constraints imposed by the Born rule.*

The functional relationship given by Eq. 11.2 depends on details of the reference measurement. In the special case that the reference measurement is a symmetric informationally complete POVM (SIC), Eq. 11.2 takes the simple form

$$p(F_j) = TrF_j\rho_i = \sum_i \left((d+1)p(E_i) - \frac{1}{d}\right) p(F_j|E_i). \qquad (11.4)$$

This is a consequence of the complete information representation of a quantum state and the use of SIC-POVMs.

---

[3] In this book, we considered the Växjö generalization of FTP only for pure states and quantum observables given by Hermitian operators (see sections 10.8, 10.14.2), but this formula can be straightforwadly generalized to POVMs and mixed states [161].

[4] However, for QBists the above generalization—to start the probability update scheme with an arbitrary POVM measurement $G = (G_j)$ and not with a SIC-POVM $E = (E_i)$–seems to be unacceptable. They are really addicted on SIC-POVMs and on completeness of information gained at the first step, information about the state, even at the price of appearance of counterfactuals.

11.3 SIC-POVMs

The simplest definition is that a SIC-POVM is determined by a system of d^2 normalized vectors (ϕ_i) (they are not orthogonal) such that

$$\langle\phi_i|\phi_j\rangle| = \frac{1}{d+1}, i \neq j. \tag{11.5}$$

The elements of the corresponding SIC-POVM (E_i) are subnormalized projectors $E_i = \frac{1}{d}\Pi_i$, where Π_i is the orthogonal projector on ϕ_i. The elements of SIC-POVM E_i determine the corresponding quantum operations (atomic instruments).

The characteristic property of SIC-POVMs, symmetry, is that the inner product in the space of operators (or $d \times d$ matrices) given by the trace is constant, i.e.,

$$TrE_iE_j = const = \frac{1}{d^2(d+1)}, i \neq j.$$

By using this equality it is easy to obtain the following representation of an arbitrary density operator ρ:

$$\rho = \sum_i \left((d+1)p(i) - \frac{1}{d}\right)\Pi_i \tag{11.6}$$

where $p(i) = TrE_i\rho$ is the probability to obtain the result i for a measurement presented by the SIC-POVM (E_i).

11.4 Comparing QBism and the Växjö Interpretation

As was pointed out, QBists have conjectured that the FTP-like form of the Born rule (11.4) may be used as a basic (and maybe unique) axiom in the derivation of quantum theory. The key question that remains is in identifying what minimal further principles must be added to the equation (11.4) for the project to be successful.

This program is identical to attempts to justify the Växjö interpretation [137, 144, 153] by deriving the complex Hilbert space formalism from the Växjö generalization of FTP (sections

10.8, 10.14.2), FTP with an interference term [131, 146, 152, 153]. This generalized FTP arises as an additive perturbation of classical FTP. The perturbation term represents the interference of probabilities which is characteristic for incompatible quantum observables.

We remind once again that the Växjö interpretation is a contextual realistic interpretation of QM. Contextuality is understood in a wider sense than typically in modern discussions on Bell's inequality, in which contextuality is reduced to joint measurement with another observable. In the Växjö interpretation, contextuality is understood in the spirit of Bohr: as dependence of the outcomes of observables on the whole experimental arrangement. In particular, violation of Bell's inequality is a consequence of the complementarity of experimental contexts corresponding to different pairs of orientations of polarization beam splitters. In some sense the Växjö interpretation is an attempt to unify the views of Einstein and Bohr. This interpretation matches with the statistical interpretation of probability.

Maybe Fuchs and Schack are right that one has to find additional axioms leading to the complete derivation of quantum formalism, as, e.g., was done by D'Ariano et al. [50, 51, 70, 71]. However, proceeding with such additional axioms does not completely match with the basic principle that the quantum formalism is just a special form of the probability update generalizing the classical Bayesian update. This principle is very attractive in both QBism and the Växjö interpretation (though they have different mathematical realizations).

Completing a generalized probability update scheme by additional operational principles diminishes the value of this scheme as the *unique fundamental principle* lying in the ground of QM. It shifts the line of research to more traditional operational approaches starting with the pioneering contribution of Heisenberg [113], then Mackey [194], and nowadays D'Ariano et al. [50, 51, 58, 59]. In this chapter we do not plan to analyze the operational axiomatics of D'Ariano et al. However, in previous derivations of QM from "natural operational principles," complex Hilbert space was always encrypted, practically explicitly, in one of the axioms (see, e.g.,

G. Mackey [194] for one of the first operational derivations of the quantum formalism).

11.5 QBism Agents: Who Are You?

In 2001 the private user's experience viewpoint on quantum *physics* made me really mad, and this was the main reason for my anti-QBism attitude [137]. However, recently I understood that the situation may be not so bad as one can imagined by reading QBism manifestos, such as those presented in section 11.2. Maybe this long way to understanding is not just my fault. QBists judge too highly the private user interpretation of QM compared with the problem of concretization of the class of such private users.

I remember that in 2001 in Växjö I asked C. Fuchs: "Suppose that your user, Ivan, lives in taiga by hunting and he has never heard about QM. Would Ivan make proper predictions about the simplest quantum experiments?" I do not remember the precise answer of C. Fuchs, but it seems it was a long story about his version of FTP (11.4), which I considered totally irrelevant to my question. Then I asked C. Fuchs the same question during the next few Växjö conferences, but did not get a satisfactory answer (at least from my viewpoint). Recently, during the 2015 Växjö conference, where QBism was heavily represented by the talks of its founders, I got a new possibility to discuss its foundations with C. Fuchs. In the after-conference email correspondence I finally got a clear answer to my old question. Of course, this answer could be found in the works of Fuchs and Schack, but it was dimmed by the strong emphasis of the private agent perspective.

So, if I understood C. Fuchs correctly, the class of agents has to be *constrained!* And the basic constraint is given by the Born rule, which is treated as an empirical rule reflecting some basic features of nature. For a moment, for us, the concrete natural basis of the Born rule is not important. It is important that QBism uses this rule as an information constraint to determine a class of, so to say "quantum agents," i.e., those who "get tickets to the QBism performance."

Thus, private users of QM are those who know the main rule of the game:

> *The probability update for quantum systems has to be done with the aid of the Born rule (or QBist version of FTP [see (11.4]).* It seems reasonable that such agents would produce reasonable predictions. Thus, Ivan from the taiga is excluded from QBist agents—finally!

The Born rule constraint is the basic necessary condition for entrance to the QBism club. At the same time it is practically a sufficient condition, because other personal characteristics of an agent making predictions about quantum experiments play subsidiary roles in relation to these predictions. Thus, in principle, one may invent an abstract (conceptual) QBism agent who makes her probabilistic predictions about the results of experiments on the basis of the Born rule. In this way QBism comes close to the recent version of the information interpretation of QM of Zeilinger–Brukner proposed by C. Brukner [42]. However, while QBists would, in principle, accept an interpretation à la Brukner, i.e., referred to as a conceptual agent, they would definitely not like to diminish the role of the private agent perspective in QBism.

11.6 QBism versus Copenhagen

By following the talks of C. Fuchs or R. Schack I always had the feeling that their views are very much in the spirit of Copenhagen. Bohr and Heisenberg always pointed out that quantum formalism is not about the "quantum physical world," but it is a representation (mathematical) of measurements performed on microsystems. Thus, from their viewpoint, QM is about knowledge (especially for Heisenberg).

Are Fuchs and his coauthors, Schack, Mermin, Caves, trying to say the same thing by just using the special interpretation of probability, the subjective probability?

I still do not have my own definite opinion about this issue. Therefore, Here, I present a long citation of D. Mermin, who knows

QBism much better than me and who claims that QBism differs crucially from all interpretations in the spirit of Copenhagen [199] (pp. 7, 8):

> A fundamental difference between QBism and any flavor of Copenhagen, is that QBism explicitly introduces each user of quantum mechanics into the story, together with the world external to that user. Since every user is different, dividing the world differently into external and internal, every application of quantum mechanics to the world must ultimately refer, if only implicitly, to a particular user. But every version of Copenhagen akes a view of the world that makes no reference to the particular user who is trying to make sense of that world.
>
> Fuchs and Schack prefer the term "agent" to "user." "Agent" serves to emphasize that the user takes actions on her world and experiences the consequences of her actions. I prefer the term user to emphasize Fuchs' and Schack's equally important point that science is a user's manual. Its purpose is to help each of us make sense of our private experience induced in us by the world outside of us.
>
> It is crucial to note from the beginning that "user" does not mean a generic body of users. It means a particular individual person, who is making use of science to bring coherence to her own private perceptions. I can be a "user." You can be a "user." But we are not jointly a user, because my internal personal experience is inaccessible to you except insofar as I attempt to represent it to you verbally, and vice-versa. Science is about the interface between the experience of any particular person and the subset of the world that is external to that particular user. This is unlike anything in any version of Copenhagen. It is central to the QBist understanding of science.

Of course, the reader can find Mermin's viewpoint on QBism as private agent (user) business does not match completely the conclusion from my discussion on the class of users belonging to the "QBism club" and constrained by the knowledge about the Born rule (see section 11.5). This is a consequence of the private user's perspective to interpreting QBism—both David Mermin and I are good friends of the founder of QBism, Christopher Fuchs, and we both got our information about QBism directly from its founder.

11.7 QBism versus the Information Interpretation of Zeilinger and Brukner

In the *information interpretation* of QM, information is the most fundamental, basic entity. Every quantized system is associated with a definite discrete amount of information (see Zeilinger [238]). This information content remains constant during the evolution of a closed system. Here a quantum state is defined in the spirit of Schrödinger [219]: *The quantum state is an expectation catalog (of probabilities for all possible outcomes).*

Here we do not see the private agent perspective to information encoded in a quantum state. Thus, in spite of some similarity—the QM formalism is about information processing—the information interpretation differs crucially from QBism by the absence of the private agent perspective. These interpretations also differ in their relation to the Copenhagen interpretation. In contrast to, e.g., Fuchs and Mermin, who distanced themselves from this interpretation (section 11.6), A. Zeilinger, who presented the basic principles of the information interpretation in 1999 [238], always emphasized its close connection with the Copenhagen interpretation; in particular, he often cited N. Bohr to emphasize a connection with Bohr's ideas. The same line of presentation continues in the joint publications of Zeilinger and Brukner [39–42], Kofler and Zeilinger [190], as well as Brukner, Zukowski, and Zeilinger [43]. Indeed, the information interpretation of QM can be considered as a modern information-theoretic version of the orthodox Copenhagen interpretation. It has some commonality with von Neumann's version of this interpretation. In particular, Zeilinger and Brukner explore heavily the concept of *irreducible quantum randomness*, which was invented by von Neumann [235].

Another key person of the information approach to QM, C. Brukner, recently published a paper on what I interpret as the universal (i.e., not private as in QBism) agent perspective on the information interpretation. I cite Brukner [42]:

> The quantum state is a representation of knowledge necessary for a hypothetical observer respecting her experimental capabilities to compute probabilities of outcomes of all possible future experiments.

Here an explicit reference to the observer's experimental capabilities is crucial, compared with my analysis of QBism in section 11.5.

It has to be noted that in this paper Brukner emphasized the closeness to QBism. With this I strongly disagree. From the QBism perspective, the wave function is in the head of a concrete private agent, e.g., in Fuchs's head, not in the head of a hypothetical observer.

11.8 Interpretations of Classical Probability Theory

11.8.1 Kolmogorov's Interpretation of Probability

In applications any mathematical formalism has to be endowed with the corresponding interpretation of its entities. Here we present the original Kolmogorov interpretation of probability (see section 11.8.2 for another interpretation of the same mathematical formalism).

Kolmogorov proposed [191] to interpret probability as follows:

> we may assume that to an event A which may or may not occur under conditions Σ, [there] is assigned a real number $P(A)$ which has the following characteristics:
>
> (a) one can be practically certain that if the complex of conditions Σ is repeated a large number of times, N, then if n be the number of occurrences of event A, the ratio n/N will differ very slightly from $P(A)$;
> (b) if $P(A)$ is very small, one can be practically certain that when conditions Σ are realized only once the event A would not occur.

Part (a) of this interpretation is nothing else than the frequency interpretation of probability; cf. with the von Mises theory and his *principle of the statistical stabilization of relative frequencies* [232–234][5] In the measure-theoretic approach this viewpoint on

[5] By this principle the probability of some concrete output α of measurement is defined as the limit of the relative frequency of realizations of α in a long (in fact, infinitely long) sequence of trials. The class of sequences of trials which can serve

probability is justified by the *law of large numbers*. However, for Kolmogorov, approximation of probability by frequencies was not the only characteristic feature of probability. Part (b) (known in foundations of probability as *Cournot's principle*) also plays an important role. This is a purely weight-type argument: if the weight assigned to an event is very small, then one can expect that such an event would never happen. We emphasize that Kolmogorov presented this weight-type argument in its strongest form: "never happens." One may proceed with a weaker form: "practically never happens."

11.8.2 Subjective Interpretation of Probability

11.8.2.1 Subjective interpretation and mathematical representation of probabilities by measures

Each scientific theory consists of the two parts, a mathematical model and an interpretation of the mathematical entities. Now we make a point: in a scientific theory the same mathematical model can have a variety of interpretations. In particular, it happened with Kolmogorov's measure-theoretic model of probability. Besides the commonly used statistical interpretation, probability measures can also be interpreted in the framework of *subjective probability theory*.

This interpretation was used by T. Bayes as the basis of his theory of probability inference (see Ramsey [214], de Finetti [63], Savage [216], Bernardo and Smith [26]). Here the probability $P(A)$ represents an *agent's personal, subjective degrees of belief* in non/occurrence of the event A. In contrast to the statistically interpreted probability, which is objective by its nature, subjective probability is by definition not objective; in other words, "it does not exist in nature" independently of an agent assigning probabilities to

---

to determine probabilities is constrained by another fundamental principle of von Mises's theory: *the principle of randomness*. Sequences of trials satisfying the latter are called *collectives* (random sequences). The principle of randomness involves the ambiguous notion of a place selection. On the one hand, the ambiguity of this notion was the main pitfall for applications of von Mises's frequency theory of probability. And nowadays it is practically forgotten. On the other hand, this ambiguity (as often happens in science) played a crucial role in establishing a modern theory of randomness and algorithms.

events. This viewpoint on probability is in direct conflict with von Mises's view of probability theory as a theory of natural phenomena, similar to, e.g., hydrodynamics. Kolmogorov and the majority of Soviet probabilists also took the active antisubjectivist position. Although Kolmogorov treated probability theory as a mathematical theory (so his viewpoint on probability theory did not coincide with Mises's viewpoint), he also interpreted it as representing an objective feature of repeatable phenomena: their statistical stability.

At the same time, since the measure-theoretic definition of Kolmogorov probability is heuristically based on the weighting-like procedure for events, it seems that the subjective interpretation matches well the mathematical framework of Kolmogorov probability spaces. Instead of assigning objective weights to events (as Kolmogorov proposed to do), subjectivists assign personal weights to events, each agent assigning his own degree of belief to an event A.

11.8.2.2 Subjective probability as the basis of classical physics?

For the fathers of QM, both Copenhagenists (as Bohr, Pauli, Dirac, von Neumann) and anti-Copenhagenists (as Einstein, de Broglie), probability was objective and statistical. Why? Why not subjective? One of the reasons for this was that all physicists learned probability starting with classical statistical mechanics and the statistical interpretation was firmly incorporated in their mind. Some of them were able to give up even causality, but not the statistical nature of probability. However, it seems that the main reason was that de Finetti's views on probability, and more generally on scientific theory, were *too revolutionary* even for the "quantum folk." The latter still wanted to have a solid objective ground: in classical world, the world of macroscopic measurement devices. But de Finetti tried to teach us that even in this macroworld, neither probability nor cause is objective; person-dependent, they have to be treated subjectively. It seems that even Copenhagenists would not accept such a position. By following de Finetti consistently, they should reconsider not only the physics of microworld, as was done in the process of creation of QM, but even the physics of macroworld.

What about QBism? QBism loudly declared that quantum probabilities have to be treated as subjective entities. For this, it was admired by some people and heavily criticized by others. It seems reasonable if QBists, by exploring the subjective interpretation of probability in QM, would say:

> **CBism** *Bayesianism has to be extended to classical statistical physics and thermodynamics!*

Unfortunately, we do not hear such a message from them—not loudly at least. Only through private communication with C. Fuchs did it become clear for me that CBism is not foreign for QBists. In particular, in 2003 C. Fuchs pointed out (see, e.g., [98], p. 812):

> Since becoming immersed in the subject, I have found nothing more exciting than these trains of thought. For they indicate the extent to which quantum foundations research may be the top of an iceberg, indeed, something with the potential to drastically change our worldview, even outside the realm of physical practice.

However, I stress once again that this message was totally shadowed by active advertising of the Bayesian subjective probability perspective solely for quantum physics. Thus, CBism still waits similar advertising and justification as QBism. And this is a program of huge importance and complexity.

11.9 QBism's Role in the Justification of Applications of Quantum Theory Outside of Physics

My personal viewpoint on the subjective interpretation of probability is sufficiently complicated. As a student of the Department of Mechanics and Mathematics at Lomonosov Moscow State University, I was lucky to have a few lectures of Kolmogorov. Then he became too ill to continue, but the lectures were given by his former student A. N. Shiryaev; in any event, "subjective probability" was a swearword for us. Therefore, by working on quantum foundations [131, 153], I always keep the statistical interpretation of probability.

However, recently by working on applications of quantum probability to cognition, psychology, and decision making [154] (see also recent papers [10, 167–169]) I started to think that the subjective interpretation is adequate for modeling the decision-making process by an individual agent. The recent wave of multi-disciplinary activity on applications of the mathematical formalism of quantum theory, and especially its probabilistic part in cognitive science, psychology, social and political sciences, economics and finances, as well as in molecular biology and genetics and modeling of biological evolution (see, e.g., [10, 45, 60, 184–186, 210, 221] and the references therein) is very supportive of QBism as a very general approach to decision making. It seems that QBism is the most natural candidate for a foundational justification of the applications of the quantum calculus of probabilities outside of physics.

Nowadays, in theory, decision-making agents are commonly considered as assigning subjective probabilities to possible actions (see Savage [216] (although the original theory of expected utility of von Neumann–Morgenstern [236] was based on the frequency interpretation).

As is well known, the use of the classical probabilistic model in decision making leads to numerous paradoxes, the most famous being the Ellsberg [76] and Machina paradoxes [193]. As was pointed out in [6, 105], these paradoxes can be resolved by using the quantum calculus of probabilities.[6] However, one has to justify the applicability of the quantum probabilistic calculus for decisions made about events in the macroworld. QBism can do this easily. The private agent perspective is not rigidly coupled to events happening in the microworld. It can be extended to decisions of agents predicting outcomes of events, e.g., at the financial market. This viewpoint on QBism as the basis of general decision making by agents processing information in accordance with the rules

[6]It is a good place to point out that these paradoxes cannot be solved simply by playing with the interpretations of probability. The essence of the problem is not an interpretation, but the structure of the probability calculus. Thus, by using QBism to resolve these paradoxes we use both its fundamental counterparts, interpretational (the subjective interpretation of probability) and mathematical (the Born rule).

formalized in the quantum probabilistic formalism was presented in [171].

Once again (cf. with the last comment of the introduction section), by extending the domain of QBism applicability to general theory of decision making, we definitely go beyond the original ideas of QBism's fathers (Fuchs, Schack, and Mermin), and it is not clear whether they would be happy to see such attempts of application of QBism, e.g., to economics or finances.

C. Fuchs and R. Schack declared [96]: "According to QBism, quantum mechanics can be applied to any physical system. QBism treats all physical systems in the same way, including atoms, beam splitters, Stern-Gerlach magnets, . . ." If so, then why only "to any physical system?" Why not to any system, biological, cognitive, social, political? From my viewpoint, QBism is an excellent interpretation to motivate extension of the domain of applications of quantum formalism. We also remark that C. Fuchs actively uses the works of W. James (see, e.g., [118–120]) as supporting the QBism perspective to decision making for outcomes of measurements. W. James was one of the great psychologists of the 19th century. In particular, he invented the principle of complementarity (conscious and unconscious thoughts) to psychology. Later N. Bohr borrowed this principle for QM. Thus, interest in the works of James might stimulate QBists to think about the interconnection between QBism and psychology, cognition.

However, it seems that, for a QBist, it is difficult (if at all possible) to accept the possibility of such wider use. The reaction of C. Fuchs to my comments in this direction cannot be characterized as excitement. Of course, this calm reaction might have social roots. By assuming that QBism is a theory about generalized probability update done not only in physics, but, in fact, everywhere, QBists would depart even farther from mainstream physics.

However, it might be that distancing from applications outside of physics has fundamental grounds. In contrast to the Växjö interpretation, in QBism the Born rule is not just a consequence of a very general scheme of the probability update. Its appearance in quantum physics is a consequence of some fundamental feature of nature, namely a kind of intrinsic quantum randomness. Thus, by extending the domain of applications of quantum formalism to cover, e.g., cognition, one has to assign to cognition a kind of intrinsic quantum(-like) randomness. In principle, one cannot exclude that not only quantum physical systems but even biosystems are intrinsically random. However, this is a very complicated problem. It seems that, for a moment, QBists (who are busy with their own problems in physics) simply do not like to be involved in the problem of justification of intrinsic bio-randomness.

However, as we know well, any idea lives its own life and sometimes its evolution may be unexpected and surprising for its creators (or even unwanted).

According to Derrida, "what we get when we read a text is not an objective account of logos or even what the author really meant, but our present interpretation or understanding of the text itself. This understanding becomes so to speak, our own [text] of the text" (quoted from [203], p. 368.)

Chapter 12

Decision Making: Quantum-Like Model of Lottery Selection

This chapter can be considered as an illustration of applications of the mathematical formalism of QM to modeling the process of decision making by humans (see section 1.3 for an introduction and general discussion). In this chapter we present a very general quantum-like model of lottery selection based on the representation of beliefs of an agent by pure quantum states. Subjective probabilities are mathematically realized in the framework of quantum probability (QP). Utility functions are borrowed from the classical decision theory. An agent processes this information by using the rules of quantum information and QP. This process is very complex; it combines counterfactual reasoning for comparison between preferences for different outcomes of lotteries which are in general complementary. These comparisons induce interference-type effects (constructive or destructive). The decision process is mathematically represented by the comparison operator and the outcome of this process is determined by the sign of the value of a corresponding quadratic form on the belief state. This operational process can be decomposed into a few subprocesses. Each of them can be formally treated as a comparison of subjective expected

Social Laser
Andrei Khrennikov

ISBN 978-981-4800-83-9 (Hardcover), 978-0-367-81720-6 (eBook)
www.jennystanford.com

utilities and interference factors (the latter express, in particular, risks related to lottery selection).

12.1 Lottery Selection: Why Quantum Probability?

As was pointed out in section 1.3, recently QP has begun to be widely used to model the processes of decision making (DM) in cognitive psychology, behavioral economics, and finance. This chapter is a compact presentation of the model designed by Basieva et al. [24] (in this paper, one can find an extensive bibliography of quantum-like modeling of decision making).

From the outset we stress that, in fact, QP is not a calculus of probabilities, but of quantum states, "complex probability amplitudes." Of course, probabilities can be generated from this calculus—via the Born rule (see section 10.3, formula (10.20)). However, intrinsically, the process of DM is represented in terms of amplitudes and not probabilities. Typically, this fundamental issue was not stressed in works on the QP-based models of DM, since the basic effects, such as violation of the Sure Thing Principle [216], were finally expressed via the Born rule in terms of probability. However, this issue is very important for our model, and we shall turn to it later.

As was remarked in section 1.3, one may consider the appeal to QP for DM modeling, instead of using conventional probabilistic measures, which are too exotic. Yet, we recall that as early as the 1970s, Tversky, Kahenman [227, 228], and other researchers in psychology and economics following the seminal paradoxes by Allais [7] and Ellsberg [76] have been demonstrating cases where classical probability (CP) prescription and actual human thinking persistently diverge, seeking to explain these deviations away from the normative DM frameworks. The main inquiry of the experimental studies was often focused on human evaluation and revision of probabilities in uncertain DM situations. The following questions naturally emerged:

> *Do people obey the rules of classical probability theory, and if not, in which circumstances? Are there any other laws that can be applied to formalize human judgments and preferences?*

After first demonstrating comprehensive evidence on deviation of human preference formation from the postulates of von Neumann and Morgenstern [236] theory (VNM), Ellsberg [76] (p. 646) proposed:

> There would be simply no way to infer meaningful probabilities for those events from their [participants] choices, and theories which purported to describe their uncertainty in terms of probabilities would be quite inapplicable in that area (unless quite different operations for measuring probability were devised).

A wide array of elegant generalizations of classical probabilistic formulations of rational decision theories (VNM) were devised following the emerging empirical evidence. We allude to von Neumann and Morgenstern's expected utility formulation [236] under objective risk, as well as Savage's subjective expected utility over consequences in uncertain states of the world [216]. Generalized utility theories focused particularly on eliminating the linearity in probabilistic measures, and seeking to relax the assumption of agents' possessing firm and state-independent probabilistic estimates.

Another stream of case-by-case explanations, based on heuristics and biases pioneered by Kahneman and Tversky [227, 228], also gained wide recognition in behavioral economics. It mainly emerged due to an extensive collection of empirical evidence on human preference formation and judgment.

Yet, these convincing explanations of each such bias and its effect on DM formation were placed under critique largely due to their lack of theoretical coherence and normative appeal. We stress that prospect theory by Kahneman and Tversky, and the advanced version, cumulative prospect theory, can be considered great accomplishments in the pursuit of a generalized and structured DM framework, encompassing some of the effects of human heuristics and biases [123–127, 227, 228].

Yet, the need for a DM framework with theoretical foundations that can be utilized in economics, finance, and other domains persists. Hence, by generalizing or replacing the classical VNM framework, one is compelled to maintain the theoretical foundations of the alternative decision theory. Probabilistic and statistical methods are undeniably the cornerstones of modern scientific methodology in all spheres of social science. Thus, although the heuristic approach to decision making cannot be discarded completely and serves as an important tool to research the nature of human reasoning, it appears that it is more natural to approach novel probabilistic models to formalize preference formation. Hence, we proceed with the slogan *QP instead of heuristics and biases!*

Application of the laws of QP (treated as the calculus of probability amplitudes), instead of CP, can resolve some paradoxes of classical DM theory (see section 12.2). The number of different paradoxes generated by the classical DM theory is startling. The authors of a recent review (Ert and Erev 2015) identified 35 basic paradoxes. The history of decision theory can be characterized by advancement of the theoretical frameworks via creation and resolution of paradoxes through modifications of the theory. As an example, the von Neumann–Morgenstern (VNM) expected utility theory was generalized to prospect theory after numerous empirical studies. However, any modification suffered from new paradoxes.

It seems that the use of QP can resolve such paradoxes (including [7], Ellsberg [76], and Machina [193] paradoxes). At least this is what Asano et al. claim in their recent paper [11] (see this paper for a detailed mathematical modeling of these paradoxes). In this paper the authors develop a model of selection of lotteries under uncertainty based on quantum formalism: representation of belief states as quantum states and introduction of *comparison operator* D, which is based on a classical utility function $u(x)$. This operator is the basic mathematical object of the theory and it provides the operational representation of the process of comparison of utilities of two lotteries. This quantum-like model can be coupled to the expected utility approach (endowed with either objective or subjective interpretations of utility functions and probabilities).

In particular, this model reproduces the VNM expected utility theory (in this case, probabilities can be interpreted objectively) and prospect theory (including its representation with cumulative probability weighting function). Moreover, the quantum-like model of lottery selection recreates *one special form of the probability weighting function used in prospect theory.* We recall that in prospect theory, the probability weighting function is an important concept to explain the violation of the independence axiom in VNM theory. Actually, from phenomenological discussions, various weighting functions have been proposed. Especially, we note the form of the two-parameter weighting function

$$w_{\lambda,\delta}(x) = \frac{\delta x^{\lambda}}{\delta x^{\lambda} + (1-x)^{\lambda}}, \tag{12.1}$$

which was discussed by Gonzalez and Wu [100]. The parameters λ and δ control the curvature and elevation of the function, respectively. Such a phenomenological function with $\lambda = 1/2$ corresponds to the subjective probabilities derived from the usage of the QP framework (see [11] for a detailed discussion). It would be interesting to find a purely psychological motivation for this specific shape of the weighting function corresponding to $\lambda = 1/2$.

The quantum-like model of Asano et al. [11] not only reproduces the output of prospect theory (for the aforementioned special choice of the weighting function) but, depending on the belief state of an agent, can lead to new decision rules, including the existence of new parameters besides the subjective probabilities. These parameters are given by relative phases expressing correlations between different outcomes of lotteries A and B, within a single lottery or between two lotteries. The presence of phases induces the effects of constructive and destructive interference.

In this chapter we advance the model proposed by Asano et al. [11] by representing Alice's beliefs about the lotteries' outcomes by two arbitrary orthonormal bases.[1] In the operator terms this corresponds to representation of beliefs by two in general noncommuting operators A and B (see (12.2)). In the article by

[1] Since the present model generalizes the model of Asano et al. [11], it resolves the paradoxes of the classical DM theory (see [11] for a detailed presentation [with corresponding numerical simulation]).

Asano et al. [11], beliefs about the lotteries were represented by commuting operators. In the case of commuting A and B the interference effects were generated solely by comparison operator D. In the general model with noncommuting lottery operators there are two sources of interference between beliefs:

(a) complementarity of lotteries
(b) reflections generated by operator D

In probabilistic terms, in the present model, the DM process is split into four sub-processes (see section 12.7). The previous model [11] handled only one of these processes (Process 1; see section 12.7). The new counterparts of the DM process model describe mathematically Alice's reflections with respect to the selection of lotteries. These reflections are modeled with the aid of *quantum transition probabilities.* For complementary lotteries (represented by noncommuting operators), these probabilities are nontrivial and their presence generates complex reflections of a decision maker; cf. [11]. The transition probabilities are involved in the creation of more complex subjective probabilities for the outcomes of lotteries than the probabilities of Process 1 only [11]. The devised model is quite complicated from the viewpoint of mathematical computations. We reproduce the detailed model derivation in section 12.10.

The structure of the model is very rich. To demonstrate at least some of its distinguishing features, we analyze in great detail the example of lotteries with two outcomes (see sections 12.8 and 12.9). This simple example shows that, in fact, a quantum-like agent uses the probability amplitudes (and not the squares of their absolute values) as weights for averaging of utility function. So, the agent evaluates expected utility via usage of square roots (probability amplitudes). Of course, the straightforward probability interpretation of this construction is impossible (since amplitudes need not be nonnegative real numbers). For the probabilistic interpretation, one has to proceed with the four counterparts of the process of decision making considered in section 12.7. At the same time, modeling based on amplitudes is attractive by its simplicity. One can proceed in this direction by using signed and complex "probabilities" which are widely used in quantum mechanics and have recently started to be applied to decision making.

Now we turn to the interpretation of QP as the calculus of probability amplitudes. In the models of Asano et al. [11] and this chaper, the decision rule is represented in the form $\langle D\Psi|\Psi\rangle \geq 0$, where Ψ is the belief state of an agent. Although, as was emphasized, this decision rule can represented in the form of comparison of averages of the utility function $u(x)$ with respect to subjective probabilities (or weighting function of them), the probabilistic representation is not intrinsic. Of course, its derivation is important to couple the quantum-like model with the existing DM models of the VNM and prospect theory type. However, such a probability representation shadowed the fundamental feature of our model, which *does not presume that agents really compare lotteries by calculation of averages (with respect to probabilities or their weighting).* An agent uses the objective or subjective probabilities (P_i) and (Q_j) corresponding to the outcomes of lotteries A and B to create her belief states about these lotteries, $|\Psi_A\rangle$ and $|\Psi_B\rangle$. Then she superposes these states to create state $|\Psi\rangle$ of her beliefs about two lotteries. This state is substituted in the quadratic form of comparison operator D and the value of this form in state of superposed beliefs Ψ determines the agent's concrete decision. This is the individual decision, not statistical. However, it can be coupled to statistics through the probabilistic decomposition of the quadratic form $\langle D\Psi|\Psi\rangle$ (see section 12.7). This decomposition expresses potentiality of realization of the concrete decisions encoded in the belied state Ψ. The comparison operator D is a cognitive feature of an agent. In this model the process of decision making is algorithmic and it is reduced to calculation of the value of the comparison form on the belief state.

We remark that, as well as prospect theory or regret theory, our quantum-like model of DM is of the descriptive type. As the creators of other descriptive theories, we try to explain paradoxes and decision problems. Our explanation is derived from the quantum theory endowed with the information interpretation of states. An agent whose information processing can be described by the quantum formalism uses the belief-weighting procedure based on coefficients in a state's expansion with respect to a basis corresponding to a set of beliefs. Such belief weighting explains the basic paradoxes of DM theory. For the moment, we do not have a neurophysiological mechanism of such weighting (see [61, 156, 180] for some attempts to model it).

12.2 Classical versus Quantum (Subjective) Expected Utility

In their book [236] von Neumann and Morgenstern (1944) introduced an expected utility function over lotteries, or gambles. The type of uncertainty which was embedded in their expected utility approach was objective uncertainty, i.e., an uncertainty which is formalizable by using objective probabilities.

We note that in economics there has been a longstanding discussion between expected value and expected utility. This debate relates to the so-called St. Petersburg paradox.

A key theorem in the VNM theory establishes the so-called expected utility representation, which in essence requires that preferences over lotteries satisfy a specified number of axioms. (In essence, those preferences need to satisfy the "substitution" and "continuity" axioms.) As economic history has shown, some of their axioms were not as "natural" as expected and the Allais paradox [7] showed a violation of the so-called substitution axiom.

This axiom says that for three lotteries, a, b, and c, the preference of lottery a over lottery b will imply that the weighted average of lotteries a and c is preferred over the weighted average over lotteries b and c.

While VNM developed an axiomatic choice framework along objective uncertainty, it surely is the case that real-life decisions can revolve around subjective choice situations. The purpose of Savage's theory is to consider choice under subjective uncertainty (see Savage [216]).

Savage [216] formulated the famous *sure-thing principle*, which is an essential axiom (among seven other axioms) that allows for the existence of an equivalence relation between a preference over acts and an ordering of expected utilities.

We remark that in purely probabilistic terms, this principle is equivalent to the validity of the law of total probability [154]. Hence, a violation of this law for our quantum-like model of DM will be equivalent to a violation of Savage's sure-thing principle (see [45, 46, 210]). We note that the well-known Ellsberg paradox [76] specifically refers to a violation of the sure-thing principle. We remark that the sure-thing principle can also be found back in the von Neumann–Morgenstern theory, but only if one considers probabilities without finite support. In the expected utility representation of the Savage approach as well as in the VNM approach, the utility function will be bounded and unique, up to an affine transformation. The linearity of the preference function is tightly connected to the substitution axiom which forms part of the VNM theory.

Violation of this axiom was shown to occur via the Allais paradox, which we already mentioned above. More paradoxes exist. The Machina paradox [193] will challenge a whole variety of expected utility approaches, such as the max–min expected utility and the Choquet expected utility.

All these DM theories are mathematically formalized with the aid of *classical probability* (CP), the axiomatics of Kolmogorov [191]. These axiomatics are based on the set representation of events and the measure representation of probabilities. Let us make the following point. The constraints posed on a DM model by the CP calculus can have fundamental consequences. The most important set of CP constraints is related to the set representation of events. In fact, this is a special representation of *classical Boolean logic.* Thus, practically all probabilistic utility models, and not only the expected utility ones, are based on the implicit assumption that all agents use the special calculus of propositions known as *Boolean algebra.* This assumption precedes, e.g. the axioms about the rationality of agents. To even formulate such axioms, one has to appeal to Boolean logic. We also remark that *expected* utility theory uses mathematical expectation which corresponds to the CP model.

It would be interesting to investigate what consequences will emerge if we consider a relaxation of some of the axioms of CP for DM under uncertainty. However, this general project has a very high complexity: one can create a huge variety of novel non-Kolmogorovean models, and to analyze all possible consequences for DM is really impossible. In particular, in mathematical applications, we can find a variety of generalized averaging procedures leading to nonclassical notions of mathematical expectation. Therefore, it would be natural to try to go beyond CP (and Boolean logic) on the basis of some concrete and well-developed non-CP model which has already demonstrated its applicability to the solution of nontrivial problems in not only natural science but also in economics, psychology, and other areas of social science.

Such a non-CP model is now well known: this is the probabilistic counterpart of the mathematical apparatus of quantum mechanics: QP. It is essential we make a remark about possible interpretations of quantum mechanics. The variety of interpretations of quantum mechanics.

Let us point to the recently developed subjective probability interpretation of QM, known as *quantum Bayesianism* (QBism) (see Chapter 11). By this interpretation, QP is the machinery used by agents to update *subjective probabilities* for outcomes of experiments. *QBism stresses the private agent perspective in quantum theory.* Development of QBism is very important for justification of applications of quantum formalism to modeling the process of decision making based on subjective probability. QBism is the only interpretation of QP based on subjective probability. Thus, QP is treated as formalizing the DM process by an agent who follows the rules of quantum logic. The latter relaxes some basic rules of classical logic. Here, in particular, an agent can violate the law of distributivity between conjunction and disjunction.

In this chapter we present a concrete QP-based model of the DM process under uncertainty which is generated by a complex information environment, including internal representations of lotteries by a decision maker. Here lotteries should not be reduced to mechanical devices such as roulettes. These are generators of events with a complex interrelation inside each of the lotteries as well as between lotteries.

12.3 Quantum Formalization of Selection of Lotteries

There are two lots, say $A = (x_i, P_i)$ and $B = (y_i, Q_i)$, where (x_i) and (y_i) are outcomes and (P_i) and (Q_i) are probabilities of these outcomes. All of the outcomes are different from each other. Which lot do you select?

12.3.1 Conventional Approach Based on Classical Probability

An agent, say Alice, can simulate the experience that she draws the lot A (or B) and gets the outcome x_i (or y_i). Let us represent such an event by (A, x_i) (or (B, x_i)). As usual, Alice assigns the utilities $u(x_i)$ and $y(x_i)$ of (A, x_i) and (B, y_i), respectively. Here, $u(x)$ is a utility function of outcome x. (In the classical model the utility of an event depends on only its outcome. However, in the quantum-like model, utility has the complex Hilbert space representation encoding all circumstances of realization of outcomes.) By using the utility function the agent evaluates various comparisons for making the preference $A \succeq B$ or $B \succeq A$.

The first mathematically consistent theory of decision making was VNM expected utility theory based on VNM axioms: completeness, transitivity, independence, and continuity. VNM axioms are given for the relation of utilities like $u \succeq v$ and the operation using probability like $pu + (1 - p)v$. This motivates an agent to operate with the expected utilities, $E_A = \sum u(x_i)P_i$ and $E_B = \sum u(y_i)Q_i$, and to use their difference as the criterion for making a preference.

However, the VNM decision theory is not free of paradoxes, as we observed in section 12.2. This problem is fundamentally coupled to the interpretation of probability used in the VNM theory. VNM used the frequency (statistical) interpretation of probability. Therefore, it is natural to test models of decision making based on other interpretations. The most powerful alternative to the frequency probability is the *subjective probability*. The subjective probability is not the frequency of an event obtained on the basis of a large number of trial experiments. Rather, it is the measure of belief about whether a specific outcome is likely to occur.

12.3.2 Belief-State Space

In our model of decision making we describe subjective probability by using the framework of quantum theory, compared with classical modeling with the aid of subjective probabilities. We emphasize that the quantum formalism operates with a state before measurement. In quantum-like models of cognition and decision making, a quantum state is treated as the belief state of an agent. Such a state formalizes the mental representation for uncertain (unmeasured) events.

Consider the space of belief states of an agent. In accordance with the quantum-like modeling of cognition, belief states are represented by normalized vectors of a complex Hilbert space H. These are so-called pure states. More generally, belief states are represented by density operators encoding classical probabilistic mixtures of pure states. But in this chapter we restrict our consideration to pure belief-states (although it will be convenient to represent them by operators.)

Lotteries A and B are mathematically realized as two orthonormal bases in H: $(|i_a\rangle)$ and $(|j_b\rangle)$ (see section 10.11 for Dirac's symbols). Any vector $|i_a\rangle$ represents the event (A, x_i): "selection of the A lottery which generates the outcome x_i." The same can be said about vectors of the B basis. These events are not real, but imaginary. Alice plays with the potential outcomes of the lotteries and compares them.

This is a kind of counterfactual reasoning. From this viewpoint, we treat quantum formalism as a mathematical device for counterfactual reasoning. Of course, we are well aware that this is not the only possible representation for such a reasoning; in future, other models of counterfactual reasoning can be in use.

We can also represent lotteries by Hermitian operators, the lotteries operators:

$$A = \sum_i x_i \; |i_a\rangle\langle i_a|, \quad B = \sum_j y_j \; |j_b\rangle\langle j_b|. \tag{12.2}$$

As in the classical theory, each outcome x_i has some utility $u_i = u(x_i)$ (say amount of money). Thus, our model is based on a mapping from eigenstates of the "lottery operators" to utilities (amounts of money), $|i_a\rangle \to u(x_i)$, $|j_b\rangle \to u(y_j)$. Thus, starting with two lotteries A and B with outcomes (x_i) and (y_j) with corresponding utilities $u_i = u(x_i)$ and $v_j = u(y_j)$, Alice represents these utilities (coupled to their lotteries) by two orthonormal bases in the belief-state space H:

$$u_i \to |i_a\rangle, \ v_j \to |j_b\rangle. \tag{12.3}$$

We emphasize coupling of utilities to lotteries. Utility (derived from some monetary amount) has not only the value but also, so to say, the "color" determined by the circumstances surrounding the corresponding lottery: *lottery's context.* Therefore, even the same outcome $z = x_i = y_j$ of two lotteries (having the same utility value $u = u(z)$) may be represented by two different vectors: $|i_a\rangle \neq |j_b\rangle$. Moreover, outcomes inside each lottery are also coupled through selection of the fixed orthonormal basis. Finally, we remark that mapping (12.3) encodes the correlations between the outcomes of lotteries (and their utilities). These correlations are mathematically expressed through quantum transition probabilities. We shall define them in the next section (see, e.g., [106, 107] for details).

The lotteries operators can be noncommuting, i.e., $[A, B] \neq 0$. In quantum theory, noncommuting Hermitian operators represent complementary (or incompatible) observables: they cannot be measured jointly. In our quantum-like model of lottery selection, we can speak about *complementary lotteries* which are represented by noncommuting operators. In the DM process for complementary lotteries, Alice does not create a joint image of the outcomes of both of them. In mathematical terms the latter means the impossibility to determine a joint probability distribution for the pairs of outcomes (x_i, y_j). Thus, instead of weighting probabilistically the pairs of outcomes, Alice analyzes the possibility of realization of an outcome, say, x_i of the A lottery and she accounts its utility $u(x_i)$. Then, under the assumption of such a realization, she imagines possible realizations (y_j) of the B lottery and compares the utilities $u(y_j)$ and $u(x_i)$.

"Suppose I have selected the A lottery and its outcome x_i was realized. What would be my earning (lost) if (instead) I were selected the B lottery and its outcome y_j were realized?"

This kind of counterfactual reflections is mathematically described by the Hilbert space formalism and transition from the A basis to the B basis. Outputs of these comparisons are weighted by accounting Hilbert space coordinates (see the discussion on QP as a calculus of probability amplitudes at the very end of introduction). This accounting is described by the special comparison operator D. Since Alice cannot handle both lotteries simultaneously, she starts with imagining one of them, say A, as in the above consideration. Then she performs similar counterfactual reasoning starting with the B lottery. The comparison operator D has two counterparts representing the processes $A \to B$ and $B \to A$ comparisons (see section 12.6, formula (12.20)). In operator terms, transitions from one basis to another are represented by transition operators $E_{i_a \to j_b}$, $E_{j_b \to i_a}$ (see (12.19)). And the comparison operator D is compounded of these operators.

12.3.3 Transition Probabilities

We now use the notion of the quantum transition (conditional) probability (see section 10.6). For our applications, it is sufficient to consider transitions between the states $(|i_a\rangle)$ and $(|m_b\rangle)$. We have (section 10.6, formula (10.41)):

$$\langle m_b | \, i_a \rangle = \sqrt{p(m_b | i_a)} e^{i\theta_{i_a \to m_b}}, \tag{12.4}$$

where $p(m_b|i_a) = p(i_a \to m_b)$ is the probability of transition from the state i_a to the state m_b. Thus,

$$p(m_b | i_a) = |\langle m_b | \, i_a \rangle|^2. \tag{12.5}$$

This is the Born rule of quantum theory. Symmetry of a scalar product implies that

$$p(i_a \to m_b) = p(m_b \to i_a), \text{ i.e., } p(m_b | i_a) = p(i_a | m_b).$$

We also remark that the corresponding transition phases are related as $\theta_{i_a \to m_b} = -\theta_{m_b \to i_a}$.

12.4 Dynamical Origin of Phases

Here we shall discuss the meaning of coefficients in the expansion of a quantum state $|\psi\rangle$ with respect to an orthonormal basis. For simplicity, we consider the two-dimensional state space. Here we represent some dichotomous observable by Hermitian operator A with the eigenvalues (x_1, x_2) and eigenvectors $|1\rangle, |2\rangle$. Any state $|\psi\rangle$ can be expanded with respect to this basis:

$$|\psi\rangle = c_1|1\rangle + c_2|2\rangle, \tag{12.6}$$

where c_1, c_2 are complex numbers and

$$|c_1|^2 + |c_1|^2 = 1. \tag{12.7}$$

By using the quantum terminology the state $|\psi\rangle$ is *superposition* of the (eigen)states $|1\rangle, |2\rangle$. We remark that the use of the linear space representation is very common in a variety of cognitive and psychological models. Thus, one might think that the only uncommon feature of the model is the use of complex numbers. However, since each complex number z can be represented as $z = u + iv$, where u, v are real numbers, any complex linear model of dimension n can be treated as a real model of dimension $2n$.

The main distinguishing feature of a quantum model is that the coefficients have a probabilistic meaning given by the famous Born rule. For the state $|\psi\rangle$ of the form (12.6), the numbers

$$p_j = |c_j|^2 \tag{12.8}$$

are interpreted as the probabilities of the outcomes x_j of the observable A having the basis of eigenvectors $|1\rangle, |2\rangle$. Thus, the absolute values of the coefficients in the expansion (12.6) have a clear meaning: these are square roots of probabilities, $|c_j| = \sqrt{p_j}$. However, any complex number also has a phase: $c_j = |c_j|e^{i\theta_j}, j = 1, 2$. The interpretation of phases is more complicated. Why do we need phases at all? Why is it not sufficient to work with states with real coefficients? From the viewpoint of the Born rule, it seems that it would be sufficient to proceed with superpositions of the form

$$|\psi\rangle = \sqrt{p_1}|1\rangle + \sqrt{p_2}|2\rangle. \tag{12.9}$$

One of the possibilities of providing a consistent interpretation to phases is to consider the *dynamical model of states generation.*

The crucial point is that to have the law of conservation of probability (see (12.7)), we have to consider *unitary dynamics.* And unitary dynamics can generate nontrivial phases starting with superpositions of the form (12.9). The dynamics of a quantum state is described by the Schrödinger equation

$$i\frac{\partial|\psi\rangle}{\partial t}(t) = \mathcal{H}|\psi\rangle(t), \ |\psi\rangle(0) = |\psi_0\rangle, \tag{12.10}$$

where $\mathcal{H}$ is the generator of quantum dynamics, a Hermitian positively definite operator. It has the dimension of frequency, i.e., 1/time.

In physics the left-hand side of the equation also contains the Planck constant, which has the dimension of action $=$ time $\times$ energy. Therefore, the generator has the dimension of energy. It is called Hamiltonian and has the meaning of energy observable. In applications outside of physics, we treat it formally as dynamics generator. In financial modeling, $\mathcal{H}$ was interpreted as a kind of financial energy [106, 107]; in social modeling it was interpreted as a kind of social energy. However, such interpretations suffer from the absence of measurement methodology.

$\mathcal{H}$ can be called an *oscillation operator*. To better understand its meaning, let us consider its eigenvalues ω_1, ω_2 and corresponding eigenvectors $|e_1\rangle$, $|e_2\rangle$. On this basis the Schrödinger equation is a system of two ordinary differential equations for the complex coordinates $z_j(t)$, $j = 1, 2$,

$$i\frac{dz_j}{dt}(t) = \omega_j z_j(t). \tag{12.11}$$

Its solution has the form

$$z_j(t) = e^{-i\omega_j t} z_{0j}. \tag{12.12}$$

These are two oscillatory processes. Their combination gives complete state oscillations:

$$|\psi\rangle(t) = e^{-i\omega_1 t} z_{01}|e_1\rangle + e^{-i\omega_2 t} z_{02}|e_2\rangle. \tag{12.13}$$

Thus even if, for the initial state, the coefficients $z_{0j} \in \mathbf{R}$, the dynamics generates nontrivial phases and complex coefficients.

Following the model of dynamical decision making [9], the agent's state evolves driven by the Schrödinger equation until the moment of decision making $T = T_{\text{dm}}$. The simplest problem of decision making can be represented as a measurement of some observable, say dichotomous, represented by a Hermitian operator A with eigenvectors $|1\rangle$, $|2\rangle$. By expanding the state $|\psi\rangle(t)$ on this basis we get the representation

$$|\psi\rangle(t) = c_1(t)|1\rangle + c_2(t)|2\rangle, \tag{12.14}$$

where $c_j(t) = e^{-i\gamma_j t}\sqrt{p_j(t)}$. Now at the instant of the self-measurement of the mental observable A an agent uses the state

$$|\psi\rangle = e^{-i\theta_1}\sqrt{p_1}|1\rangle + e^{-i\theta_2}\sqrt{p_2}|2\rangle, \tag{12.15}$$

where $\theta_j = \gamma_j T$ and $p_j = p_j(T)$. If the operator A coincides with $\mathcal{H}$, then $\theta_j = \omega_j T$ (but this is a very special situation).

For this state, an agent makes the A observation and she obtains the output a_j with the probability p_j. This is *the objective probability model of decision making*: for a large ensemble of agents the probability frequency of the output x_j equals to p_j. Another model is based on *the subjective interpretation of probability*. An agent assigns subjective probabilities of the outputs by extracting them from the state (12.15), then she computes odds $O(1/2) = \frac{p_1}{p_2}$ and makes her choice depending on the value of odds.

In this chapter we study a more complex problem of the comparison of two lotteries which cannot be reduced to quantum-like modeling of a single observable. The process of comparison involves two, in general, incompatible observables A and B. We shall proceed with the subjective interpretation of probabilities. However, the main feature of the quantum-like process of decision making, namely, reduction of this process to elementary oscillations, will be crucial even in the coming model of comparison of lotteries (see section 12.7).

12.5 Belief State of a Decision Maker

The state of Alice's beliefs about the lottery A can be represented as superposition

$$|\Psi_A\rangle = \sum_i \sqrt{P_i} e^{i\theta_{ai}} |i_a\rangle.$$

The probability of realization of the event (A, x_i) is given by the Born rule and equals to $P_i = |\langle i_a|\Psi_A\rangle|^2$. In the same way the state of beliefs about the lottery B can be represented as superposition

$$|\Psi_B\rangle = \sum_i \sqrt{Q_i} e^{i\theta_{bi}} |i_b\rangle.$$

To point that an index serves to describe the lottery A (lottery B), we shall label it by its own index, say i_a or j_b. And we omit these labels, a, b, when the meaning of indexes is clear or their coupling to A and B is not important.

Alice superposes her belief states about the lotteries, and her total belief state is created via superposition of her beliefs about the A lottery and the B lottery. Thus, the overall ψ is the superposition of ψ's for two individual lotteries.

$$|\Psi\rangle = |\Psi_A\rangle + |\Psi_B\rangle. \tag{12.16}$$

However, since in general the states representing Alice's beliefs about the lotteries are not orthogonal,[2] i.e., in general $\langle\Psi_A|\Psi_B\rangle \neq 0$, the vector $|\Psi\rangle$ is not normalized and the state of combined beliefs is obtained via normalization: $|\Phi\rangle = |\Psi\rangle/\||\Psi\rangle\|$. To make presentation simpler, we shall proceed with the vector $|\Psi\rangle$, i.e., without normalization. Since the normalization factor is positive and the decision rule is based on inequality $\langle D\Psi|\Psi\rangle \geq 0$, this factor does not play any role in the process of comparison of lotteries (see section 12.6). In future we shall call such vectors *unnormalized states or simply states* (if this would not lead to misunderstanding).

In further calculations it is useful to use the operator representation of $|\Psi\rangle$:

$$\sigma \equiv \sigma_\Psi = |\Psi\rangle\langle\Psi| = \sigma_A + \sigma_B + \sigma_{B\to A} + \sigma_{A\to B}, \tag{12.17}$$

where

$$\sigma_A = |\Psi_A\rangle\langle\Psi_A| = \sum_{i,j} \sqrt{P_i P_j} e^{i(\theta_{ai}-\theta_{aj})} |i_a\rangle\langle j_a|$$

$$\sigma_B = |\Psi_B\rangle\langle\Psi_B| = \sum_{i,j} \sqrt{Q_i Q_j} e^{i(\theta_{bi}-\theta_{bj})} |i_b\rangle\langle j_b|$$

[2] Nonorthogonality of the states $|\Psi_A\rangle$ and $|\Psi_B\rangle$ means that the beliefs about two lotteries are not complementary. There is an "overlap" between them. The presence of such an overlap plays an important role in the process of decision making (see section 12.8).

$$\sigma_{B\to A} = |\Psi_A\rangle\langle\Psi_B| = \sum_{i,j}\sqrt{P_i Q_j}e^{i(\theta_{ai}-\theta_{bj})}|i_a\rangle\langle j_b|$$

$$\sigma_{A\to B} = |\Psi_B\rangle\langle\Psi_A| = \sum_{i,j}\sqrt{P_i Q_j}e^{-i(\theta_{ai}-\theta_{bj})}|j_b\rangle\langle i_a|.$$

We remark that, since $|\Psi\rangle$ is not normalized, $\mathrm{tr}\sigma \neq 1$. But we repeat that this is not important in the process of lotteries selection described in section 12.6.

If we consider the normalized state $|\Phi\rangle$, then the corresponding operator $\sigma_\Phi = |\Phi\rangle\langle\Phi|$ is a density operator, the operator representation of pure state $|\Phi\rangle$. But we stress that the state of beliefs about two lotteries is a pure state.

The ability to create superposition of states is the main distinguishing feature of an agent using the rules of quantum logic. However, besides creation of superpositions, the quantum-like Alice can also create mixtures of belief states:

$$\sigma = \frac{1}{2}[\sigma_A + \sigma_B]. \tag{12.18}$$

By operating with such mixtures, Alice can reproduce decision making corresponding to the VNM expected utility model.

12.6 Operator Representation of the Process of Comparison of Lotteries

In the classical expected utility theory, Alice calculates the averages of the utility function. In the quantum-like mode presented by Asano et al. [11], the utility function determines the *comparison operator*. Invention of such an operator is based on mappings from eigenstates of the "lottery operators" to utilities (amounts of money), $|i_a\rangle \to u(x_i)$ and $|j_b\rangle \to u(y_j)$ (see section 12.3.2).

We remark that we simply borrow the utility function from classical (objective or subjective) utility theory. Thus, we do not contribute, e.g., to the analysis of possible shapes of utility functions. Then we use QP to model subjective probabilities. And this modeling is encoded in the quantum representation of the belief state. However, this representation is only the first step towards the quantum world of decision making. The crucial step is the quantum-like operational description of the process of comparison of lotteries with the aid of quantum state transitions which are encoded in the comparison operator. This process can be structured as a combination of comparison of a few SEUs and the interference-type factors of the $\cos\theta$ form, where θ represents the combination of the phases of a few processes of the preferred outcomes of the lotteries. These interference factors represent additional features of the subjective images of lotteries in the mind of an agent, in particular, *risks* (see [11] for a discussion).

Let us introduce the *transition operators*

$$E_{i_a \to j_b} = |j_b\rangle\langle i_a|, \; E_{j_b \to i_a} = |i_a\rangle\langle j_b|. \tag{12.19}$$

We have, e.g., $E_{i_a \to j_b}|i_a\rangle = |j_b\rangle$. This operator describes the process of transition from preferring the state $|i_a\rangle$ to preferring the state $|j_b\rangle$. The operator $E_{j_b \to i_a} = |i_a\rangle\langle j_b|$ describes transition in the opposite direction. We stress that these are transitions between the belief states of Alice. We remark that $E_{j_b \to i_a} = E^{\star}_{i_a \to j_b}$, i.e., elementary transitions in opposite directions are represented by adjoint operators.

Now we introduce the two comparison operators

$$\begin{aligned}
D_{B\to A} &= \sum_{n,m} (u(x_n) - u(y_m)) e^{i\gamma_{m_b \to n_a}} E_{m_b \to n_a} \\
&= \sum_{n,m} (u(x_n) - u(y_m)) e^{i\gamma_{m_b \to n_a}} |n_a\rangle\langle m_b|, \\
D_{A\to B} &= \sum_{n,m} (u(y_m) - u(x_n)) e^{i\gamma_{n_a \to m_b}} E_{n_a \to m_b} \\
&= \sum_{n,m} (u(y_m) - u(x_n)) e^{i\gamma_{n_a \to m_b}} |m_b\rangle\langle n_a|.
\end{aligned}$$

The operator $D_{B\to A}$ represents the utility of selection of the lottery A relative to the utility of selection of the lottery B. We can say that by transition from the potential outcome (B, y_m) to the potential outcome (A, x_n) Alice earns utility $u(x_n)$ and at the same time she loses utility $u(y_m)$. (If $u(x) = x$ and x has the meaning of cash amounts (say US dollars), then by such a transition Alice (potentially) earns $x_n - y_m$ US dollars.)

In the same way we interpret the transition operator $D_{A\to B}$. This operator represents the utility of selection of the lottery B

relative to the utility of selection of the lottery A. These operators represent the process of Alice's reflections in the process of decision making. Her mind fluctuates between preferring outcomes of the A lottery to outcomes of the B lottery (formally represented by the operator $D_{B\to A}$) and inverse-preferring (formally represented by the operator $D_{A\to B}$). Finally, she has to compare how much she can earn (in average) by preferring A to B compared with preferring B to A. This process is formally described by *the complete comparison operator:*

$$D = D_{B\to A} - D_{A\to B}. \tag{12.20}$$

The operators $D_{B\to A}$ and $D_{A\to B}$ are basic for comparison of lotteries. As everywhere in quantum theory, the order of preferring plays a crucial role. The first operator encodes preferring of A over B and the second one encodes preferring of B over A. Then, Alice compares their outputs. This last comparison is encoded in the operator D.

Operator D has the form

$$D = \sum_{n,m}(u(x_n) - u(y_m))e^{i\gamma_{m_b\to n_a}}|n_a\rangle\langle m_b| \tag{12.21}$$

$$-\sum_{n,m}(u(y_m) - u(x_n))e^{i\gamma_{n_a\to m_b}}|m_b\rangle\langle n_a|$$

$$= \sum_{n,m} u_{nm}(e^{i\gamma_{m_b\to n_a}}|n_a\rangle\langle m_b| + e^{i\gamma_{n_a\to m_b}}|m_b\rangle\langle n_a|),$$

where

$$u_{nm} = u(x_n) - u(y_m).$$

Since all quantum observables are represented by Hermitian operators, the phases should be related as follows:

$$\gamma_{n_a\to m_b} = -\gamma_{m_b\to n_a}. \tag{12.22}$$

The comparison operator D gives us integral judgment. Only heuristically can we treat the D-based judgment as a result of the comparison of two relative utilities represented by the operators $D_{B\to A}$ and $D_{A\to B}$. We remark that the operators $D_{B\to A}$ and $D_{A\to B}$ are not Hermitian. Hence, they cannot be treated as observables. We have that $D^\star_{A\to B} = -D_{B\to A}$ and $D = D_{B\to A} + D^\star_{B\to A}$.

The quantum analogue of (subjective) expected utility theory is based on the natural decision rule:

Decision rule. *If the average of the comparison operator D is nonnegative, i.e.,* $\langle D\rangle = \operatorname{tr} D\sigma = \langle D\Psi|\Psi\rangle \geq 0$, *then* $A \succeq B$.

Using Eqs. 12.17 and 12.20 the trace can be written as the sum of four components:

$$\operatorname{tr} D\sigma = \frac{1}{2}\operatorname{tr} D\sigma_A + \frac{1}{2}\operatorname{tr} D\sigma_B + \Delta_1 + \Delta_2,$$

where

$$\Delta_1 = \frac{1}{2}\operatorname{tr}(D_{B\to A}\sigma_{A\to B} - \operatorname{tr} D_{A\to B}\sigma_{B\to A}),$$

$$\Delta_2 = \frac{1}{2}\operatorname{tr}(D_{B\to A}\sigma_{B\to A} - \operatorname{tr} D_{A\to B}\sigma_{A\to B}).$$

Which, together with the formal expressions for each components derived in section 12.10, facilitates an operational understanding of the decision-making process, as we shall see in the next section.

From calculations in section 12.10, we obtain that

$$\frac{1}{2}\operatorname{tr} D\sigma_A = \sum_{i,j,m} u_{jm}\sqrt{p(m_b|i_a)P_i P_j}\cos\Theta^A_{ij;m}, \tag{12.23}$$

where $\Theta^A_{ij;m} = \theta_{i_a\to m_b} - \gamma_{i_a\to m_b} + \theta_{ai} - \theta_{aj}$.

$$\frac{1}{2}\operatorname{tr} D\sigma_B = \sum_{i,j,n} u_{nj}\sqrt{p(n_a|i_b)Q_i Q_j}\cos\Theta^B_{ij;n} \tag{12.24}$$

where $\Theta^B_{ij;n} = \theta_{i_b\to n_a} - \gamma_{i_b\to n_a} + \theta_{bi} - \theta_{bj}$.

Calculations in section 12.10 show that it is natural to consider the following combinations of traces for comparison operators and "transition states" (see (12.51), (12.52)):

$$\Delta_1 = \frac{1}{2}\operatorname{tr}(D_{B\to A}\sigma_{A\to B} - D_{A\to B}\sigma_{B\to A}) = \sum_{i,j} u_{ij}\sqrt{P_i Q_j}\cos\Theta_{ij}, \tag{12.25}$$

where $\Theta_{ij} = \theta_{bj} - \theta_{ai} + \gamma_{j_b\to i_a}$;

$$\Delta_2 = \frac{1}{2}\operatorname{tr}(D_{B\to A}\sigma_{B\to A} - D_{A\to B}\sigma_{A\to B}) \tag{12.26}$$
$$= \sum_{i,j,n,m} u_{nm}\sqrt{p(m_b|i_a)p(n_a|j_b)P_i Q_j}\cos\Gamma_{ij,nm},$$

where $\Gamma_{ij,nm} = \theta_{j_b\to n_a} - \theta_{i_a\to m_b} + \gamma_{n_a\to m_b} + \theta_{bj} - \theta_{ai}$.

12.7 Analysis of Operator-Based Comparison of Lotteries

As we have seen, the average of the comparison operator is naturally decomposed into four counterparts representing special subprocesses of the process of decision making. We start with the simplest expression.

Process 1: Its output is represented by the quantity Δ_1, see (12.25). To simplify considerations, let us assume that all phases Θ_{ij} in the sum are equal, i.e., $\Theta_{ij} \equiv \Theta$. Thus,

$$\Delta_1 = \left[\sum_i u(x_i)\sqrt{P_i}\sum_j \sqrt{Q_j} - \sum_j u(y_j)\sqrt{Q_j}\sum_i \sqrt{P_i}\right] \cos\Theta.$$

(As was mentioned, the model has a very rich structure. It can generate very complex behavioral patterns. We consider only the simplest of them.)

Following Asano et al. [11], consider the normalized difference

$$\frac{\Delta_1}{\sum_{i,j}\sqrt{P_i Q_j}} = \left[\sum_i u(x_i)\tilde{P}_i - \sum_j u(y_j)\tilde{Q}_j\right] \cos\Theta, \qquad (12.27)$$

where the quantities

$$\tilde{P}_i = \frac{\sqrt{P_i}}{\sum_i \sqrt{P_i}}, \quad \tilde{Q}_j = \frac{\sqrt{Q_j}}{\sum_j \sqrt{Q_j}} \qquad (12.28)$$

can be interpreted as subjective probabilities. We remark that $\sum_i \tilde{P}_i = 1$ and $\sum_j \tilde{Q}_j = 1$. Alice assigns these probabilities to outcomes of the lotteries in the process of comparison of the relative utility of the $B \to A$ transition (based on her beliefs encoded in the $\sigma_{B\to A}$ counterpart of her belief state) with the relative utility of the $A \to B$ transition (based on her beliefs encoded in the $\sigma_{A\to B}$ counterpart of her belief state). We remark that the quantities

$$\langle u\rangle_{\tilde{P}} = \sum_i u(x_i)\tilde{P}_i \text{ and } \langle u\rangle_{\tilde{Q}} = \sum_j u(y_j)\tilde{Q}_j$$

are *expected utilities* for the lotteries with respect to these subjective probabilities. Thus, for this part of the process of comparison of two lotteries, Alice assigns subjective probabilities $(\tilde{P}_i)$ and $(\tilde{Q}_j)$, given by the square root transformation of the original probabilities

(P_i) and (Q_j). (The latter can be treated as objective statistical probabilities. As was pointed out in the introduction, quantum formalism specifies the square root transformation as the map determining subjective probabilities. This is a very special form of the weighting function (see (12.1)) used in the classical SEU theory. As already pointed out, it is important to find a purely psychological justification for fixing $\lambda = 1/2$ in (12.1).)

Then, she calculates subjective expected utilities and compares them. The final step of comparison takes into account the sign of the factor $\cos\Theta$. This is really a nontrivial quantum counterpart of the decision process.

We remark that the square root transformation of probabilities can be directly coupled to a selection of weighting functions in *prospect theory* (see (12.1) in the introduction).

Process 2: Now we analyze the counterpart of the decision-making process based on the comparison average given by the quantity

$$\frac{1}{2}\mathrm{tr}\ D\sigma_A = \sum_{i,j,m} u_{jm}\sqrt{p(m|i)P_iP_j}\cos\Theta_{ij;m}.$$

See (12.23). In this subprocess Alice uses only the part of her belief state (given by Ψ_A) representing her beliefs about the A lottery. She compares these A beliefs with possible transitions to the B states. Such transitions are expressed through the transition probabilities $p(m|i) = p(i_a \to m_b)$ and the phases $\theta_{i_a\to m_b}$ and $\gamma_{i_a\to m_b}$. These phases represent correlations between the events (A, x_i) and (B, x_m). These quantities are of subjective nature. Alice tries to treat the two lotteries separately, but she has a variety of correlations between them coming from the analysis of the situation and the previous experience. The transition probabilities $p(m|i)$ are also subjective quantities.

To simplify analysis, we again assume that all transition phases $\Theta_{ij;m}$ are equal, i.e., $\Theta_A \equiv \Theta_{ij;m}$. Besides the subjective probabilities $\tilde{P}_i$, we consider the subjective probabilities of transition from the lottery A to the lottery B given by

$$\tilde{Q}_{m;A} = \frac{\sum_i \sqrt{p(m|i)P_i}}{\sum_{i,m}\sqrt{p(m|i)P_i}}. \tag{12.29}$$

(We have $\sum_m \tilde{Q}_{m;A} = 1$.)[3] Consider now the normalized quantity

$$\frac{\mathrm{tr} D\sigma_A}{2\sum_{i,j,m}\sqrt{p(m|i)P_i P_j}} = \left[\sum_j u(x_j)\tilde{P}_j - \sum_m u(y_m)\tilde{Q}_{m;A}\right]\cos\Theta_A$$
$$= [\langle u\rangle_{\tilde{P}} - \langle u\rangle_{\tilde{Q}_A}]\cos\Theta_A,$$

where the quantities in the brackets are expected utilities with respect to the corresponding subjective probability distributions. Thus, in this decision subprocess, Alice assigns subjective probabilities to the A events (by using the square root transformation). Then she assigns subjective probabilities $p(m|i)$ for transitions $i_a \to m_b$. They can be interpreted in the following way: Alice assumed that the event (A, x_i) would happen (with the probability $\tilde{P}_i$, but with the probability $p(m|i)$ she changes her mind and assumes that the event (B, y_m) would happen. The output of such mental fluctuations is assignment of subjective probability to the events (B, y_m) conditioned on the beliefs about the A states, the probability $\tilde{Q}_{m;A}$. Then, Alice computes the difference between expected utilities. If $\cos\Theta_A \geq 0$, then she uses this difference to order the lotteries as $A \succeq B$. However, if $\cos\Theta_A \leq 0$, then $B \succeq A$. Thus, the ordering is opposite to SEU ordering. (We recall that this is not the final ordering of the lotteries, but just ordering generated by the subprocess under consideration.) From the viewpoint of classical SEU theory, this situation seems to be counterintuitive. The role of phases in quantum-like modeling of lottery selection is a complex question. It was discussed by Asano et al. [11], who associated the $\cos\theta$ factor with risks.

Process 3: The counterpart of the decision process based on a comparison corresponding to Alice's beliefs about the B lottery is analyzed similarly—again under the assumption of coincidence of all phases: in (12.24) $\Theta_B \equiv \Theta_{ij;n}$. Consider subjective probabilities of transition from lottery B to lottery A given by

$$\tilde{P}_{n;B} = \frac{\sum_i \sqrt{p(n|i)Q_i}}{\sum_{i,n}\sqrt{p(n|i)Q_i}}. \tag{12.30}$$

---

[3] If Alice were acting on the basis of the classical (Kolmogorov) probability and if she were not using the square root weighting of probabilities, then the quantity $\tilde{Q}_{m;A}$ would be equal to the original probability $Q_m = \sum_i p(m|i)P_i$.

(We have $\sum_m \tilde{P}_{n;B} = 1$.) Then we have

$$\frac{\mathrm{tr} D\sigma_B}{2\sum_{i,n}\sqrt{p(n|i)Q_i Q_j}} = \left[\sum_n u(x_n)\tilde{P}_{n;B} - \sum_j u(y_j)\tilde{Q}_j\right]\cos\Theta_B \tag{12.31}$$

$$= [\langle u\rangle_{\tilde{P}_B} - \langle u\rangle_{\tilde{P}}]\cos\Theta_B.$$

Thus, the output of this counterpart of the decision process is based on a comparison of subjective expected utilities and relative phases. One of the expected utilities is based on the subjective account of probability of the transition (in Alice's mind) from the beliefs about B lottery to A lottery and another is, so to say, straightforward subjective probability based on the square root transform of the initial probabilities (Q_j).

Process 4: Finally, we analyze the most complicated counterpart of the process of decision making corresponding to the comparison term Δ_2. This term compares the utilities of transitions $B \to A$ and $A \to B$ when Alice appeals to the counterparts of her state representing the beliefs about these transitions. This process is characterized by ambiguity and intensive fluctuations of Alice's mind in both directions; the intensity of these fluctuations is given by the transition probabilities $p(n_a|j_b) = p(j_b \to n_a)$ and $p(m_b|i_a) = p(i_a \to m_b)$. Correlations between outcomes of lotteries are encoded in the phases $\theta_{j_b \to n_a}$ and $\theta_{i_a \to m_b} = -\theta_{m_b \to i_a}$. Since the general form of the dependence of Δ_2 on the phases is very complex, we again assume that

$$\Delta_2 = \sum_{i,j,n,m} u_{nm}\sqrt{p(m_b|i_a)p(n_a|j_b)P_i Q_j}\cos\Gamma, \tag{12.32}$$

where $\Gamma = \theta_{j_b \to n_a} - \theta_{i_a \to m_b} + \gamma_{n_a \to m_b} + \theta_{bj} - \theta_{ai}$. Now, as in the previous processes, we can represent this comparison term as the difference of two expected utilities with respect to the subjective probabilities $(\tilde{P}_{i;B})$ and $(\tilde{Q}_{j;A})$:

$$\frac{\Delta_2}{\sum_{i,j,n,m}\sqrt{p(m_b|i_a)p(n_a|j_b)P_i Q_j}}\cos\Gamma$$

$$= \left[\sum_n u(x_n)\frac{\sum_i\sqrt{p(n|j)Q_j}}{\sum_{j,n}\sqrt{p(n|j)Q_j}} - \sum_m u(y_m)\frac{\sum_i\sqrt{p(m|i)P_i}}{\sum_{i,m}\sqrt{p(m|i)P_i}}\right]\cos\Gamma$$

$$= [\langle u\rangle_{\tilde{Q}_A} - \langle u\rangle_{\tilde{P}_B}]\cos\Gamma.$$

To finalize this the most complex counterpart of the process of comparison of the lotteries, Alice takes into account the signs of difference between subjective expected utilities and of the interference cos term.

Complete process of lottery selection. Under the simplification assumption about phases, we can write the average of the comparison operator as

$$\langle D\rangle_\Psi = c_1[\langle u\rangle_{\tilde{P}} - \langle u\rangle_{\tilde{Q}}]\cos\Theta + c_2[\langle u\rangle_{\tilde{P}} - \langle u\rangle_{\tilde{Q}_A}]\cos\Theta_A \\ + c_3[\langle u\rangle_{\tilde{P}_B} - \langle u\rangle_{\tilde{P}}]\cos\Theta_B + c_4[\langle u\rangle_{\tilde{Q}_A} - \langle u\rangle_{\tilde{P}_B}]\cos\Gamma, \tag{12.33}$$

where the weights $c_j > 0$, $j = 1, 2, 3, 4$, can be found in the above considerations for the subprocesses 1–4.

In the accordance with the decision rule, if $\langle D\rangle \geq 0$, then $A \succeq B$.

12.8 Lotteries with Two Outcomes: Uniform Probability Distribution

Consider two lotteries, A and B, having two outcomes, (x_1, x_2) and (y_1, y_2), and the utilities $u_i = u(x_i)$ and $v_j = u(y_j)$. The probabilities P_i, Q_j will be specified later. We start with calculation of the matrix of the comparison operator D in the basis $|1_a\rangle$, $|2_a\rangle$. By definition we have

$$D = (u_1 - v_1)[|1_a\rangle\langle 1_b| + |1_b\rangle\langle 1_a|] + (u_1 - v_2)[|1_a\rangle\langle 2_b| + |2_b\rangle\langle 1_a|] \\ +(u_2 - v_1)[|2_a\rangle\langle 1_b| + |1_b\rangle\langle 2_a|] + (u_2 - v_2)[|2_a\rangle\langle 2_b| + |2_b\rangle\langle 2_a|].$$

Consider in the qubit space two bases

$$|1_a\rangle = \begin{pmatrix} 1 \\ 0 \end{pmatrix}, \qquad |2_a\rangle = \begin{pmatrix} 0 \\ 1 \end{pmatrix}, \tag{12.34}$$

and

$$|1_b\rangle = (|1_a\rangle + |2_a\rangle)/\sqrt{2} = \frac{1}{\sqrt{2}}\begin{pmatrix} 1 \\ 1 \end{pmatrix}, \\ |2_b\rangle = (|1_a\rangle - |2_a\rangle)/\sqrt{2} = \frac{1}{\sqrt{2}}\begin{pmatrix} 1 \\ -1 \end{pmatrix}. \tag{12.35}$$

We can find the matrix of the comparison operator D in the basis $|1_a\rangle, |2_a\rangle$:

$$D = \frac{1}{\sqrt{2}} \begin{pmatrix} 4u_1 - 2v_1 - 2v_2 & 2u_2 - 2v_1 \\ 2u_2 - 2v_1 & 2v_2 - 2v_1 \end{pmatrix}$$

Let now $P_1 = P_2 = Q_1 = Q_2 = 1/2$, i.e.,

$$|\psi_A\rangle = (|1_a\rangle + |2_a\rangle)/\sqrt{2} = \frac{1}{\sqrt{2}} \begin{pmatrix} 1 \\ 1 \end{pmatrix},$$

$$|\psi_B\rangle = (|1_b\rangle + |2_b\rangle)/\sqrt{2} = \begin{pmatrix} 1 \\ 0 \end{pmatrix}. \tag{12.36}$$

$$|\psi\rangle = |\psi_A\rangle + |\psi_B\rangle = \frac{1}{\sqrt{2}} \begin{pmatrix} 1+\sqrt{2} \\ 1 \end{pmatrix}. \tag{12.37}$$

We obtain the following inequality determining selection of lotteries:

$$\begin{aligned} \langle\psi|D|\psi\rangle &= C\Big((1+\sqrt{2})^2 u_1 + (1+\sqrt{2}+1)u_2 \\ &\quad -((1+\sqrt{2})^2 + 2(1+\sqrt{2}) + 1)v_1 - ((1+\sqrt{2})^2 - 1)v_2\Big) \\ &= C'\Big([(1+\sqrt{2})u_1 + u_2] - [(1+\sqrt{2})v_1 + v_2]\Big) \geq 0, \end{aligned} \tag{12.38}$$

where C and C' are positive factors.

The main point is that the weights of utilities for both lotteries coincide with the coefficients (up to normalization) in the expansions of the (unnormalized) state $|\psi\rangle$ (which combines Alice's beliefs about the lotteries) with respect to the corresponding bases. We can rewrite the comparison inequality as follows:

$$(u_1 \tilde{P}_1 + u_2 \tilde{P}_2) \geq (v_1 \tilde{Q}_1 + v_2 \tilde{Q}_2), \tag{12.39}$$

where the subjective probabilities are given by the expressions

$$\tilde{P}_1 = \tilde{Q}_1 = \frac{(1+\sqrt{2})}{2+\sqrt{2}} = \frac{1}{\sqrt{2}},$$

$$\tilde{P}_2 = \tilde{Q}_2 = \frac{1}{2+\sqrt{2}}.$$

Thus, by using the state expansion with respect to the A basis (see (12.34)), and by *treating amplitudes, the coefficients with respect to the A basis, as subjective probabilities*, Alice calculates SEU for the

A lottery. We stress that she uses the subjective "probabilities" and "amplitudes" encoded in the state $|\psi\rangle$ combining beliefs about both lotteries and not probabilities (objective or subjective) encoded in the state $|\psi_A\rangle$ representing solely beliefs about the *A* lottery. Now we remark that the complete belief state $|\psi\rangle$ can be represented as well in the form

$$|\psi\rangle = (1 + \sqrt{2})|1_b\rangle + |2_b\rangle. \qquad (12.40)$$

Alice uses this representation to calculate SEU for the *B* lottery.

How can one explain, e.g., the increase of probability of the outcome x_1 compared with the outcome x_2? In the state $|\psi_A\rangle$ both these outcomes are equally possible. Now Alice started to combine (by using the rules of quantum logic)[4] the beliefs about the two lotteries, and she found that some beliefs about the *B* lottery encoded in the state $|\psi_B\rangle$ can be interpreted as the beliefs in favor of the outcomes x_1 and x_2, but the additional weight assigned to x_1 is higher than the weight assigned additionally to x_2. (In our example the latter is zero.) This is a kind of *constructive and destructive interference* for probabilities assigned to beliefs in favor of the outcomes x_1 and x_2, respectively. We recall that

$$|\psi_B\rangle = (|1_b\rangle + |2_b\rangle)/\sqrt{2} = \frac{1}{2}[(|1_a\rangle + |2_a\rangle) + (|1_a\rangle - |2_a\rangle)]. \qquad (12.41)$$

The beliefs in favor of x_1 which are present in the states $|1_b\rangle$ and $|2_b\rangle$ interfere constructively and the beliefs in favor of x_2 interfere destructively.

In this simple two-dimensional example we considered the quantum process of lottery selection straightforwardly as weighting of the utility function with respect to the coefficients in the expansion of the complete belief state with respect to the lotteries' bases and comparison of such weighted sums—"averages." The example is good for illustrative purposes, since here the coefficients are positive. This gives the possibility of interpreting them straightforwardly as probabilities and the weighted sums as averages of SEU. In general, such direct coupling to subjective probability is impossible (see section 12.9), where we consider the

---

[4]These rules are formally represented by linear algebra in complex Hilbert space. So, Alice expands $|\psi_B\rangle$ with respect to the *A* basis.

case of lotteries with arbitrary objective probabilities P_1, P_2 and Q_1, Q_2. As was shown in section 12.7, in general such coupling can be established through decomposition of the process of decision making into four subprocess. Each of them can be described in terms SEU with QP realization of subjective probabilities. The example of this section can (but need not) be considered as a sign that it may be that decision makers do not use the language of probability (neither objective nor subjective) at all, but they proceed with signed or even complex weighting of utility (see section 12.9) for a continuation of this discussion.

12.9 Lotteries with Two Outcomes: General Case

We consider the same bases for the lotteries as in section 12.8 (see (12.34) and (12.35)), but now the probabilities P_1, P_2 and Q_1, Q_2 are arbitrary.

$$|\psi_A\rangle = \sqrt{P_1}|1_a\rangle + \sqrt{P_2}|2_a\rangle, \; |\psi_B\rangle = \sqrt{Q_1}|1_b\rangle + \sqrt{Q_2}|2_b\rangle. \tag{12.42}$$

The complete belief state can be written as

$$|\psi\rangle = c_1|1_a\rangle + c_2|2_a\rangle, \tag{12.43}$$

where

$$c_1 = \sqrt{P_1} + \frac{\sqrt{Q_1}+\sqrt{Q_2}}{\sqrt{2}}, \; c_2 = \sqrt{P_2} + \frac{\sqrt{Q_1}-\sqrt{Q_2}}{\sqrt{2}}. \tag{12.44}$$

Hence, the decision inequality has the form

$$\langle\psi|D|\psi\rangle \sim \left([2c_1^2u_1 + 2c_1c_2u_2] - [(c_1+c_2)^2v_1 + (c_1^2 - c_2^2)v_2]\right) \geq 0. \tag{12.45}$$

It can be rewritten as

$$\langle\psi|D|\psi\rangle \sim 2c_1[c_1u_1 + c_2u_2] - (c_1+c_2)\,[(c_1+c_2)v_1 + (c_1-c_2)v_2] \geq 0. \tag{12.46}$$

Thus Alice assigns to the outcomes of the lotteries the weights c_1, c_2 and $d_1 = c_1 + c_2$, $d_2 = c_1 - c_2$. As we have seen in Eq. 12.43, the weights c_1, c_2 correspond to the coefficients in the expansion of the complete state $|\psi\rangle$ with respect to the A basis. It is easy to see that

$$|\psi\rangle = \frac{d_1|1_b\rangle + d_2|2_b\rangle}{\|d_1|1_b\rangle + d_2|2_b\rangle\|}. \tag{12.47}$$

The comparison inequality (12.46) can be written as comparison of two subjective utilities with respect to the probabilities

$$\tilde{P}_1 = \frac{c_1}{c_1 + c_2} = \frac{\sqrt{P_1} + \frac{\sqrt{Q_1}+\sqrt{Q_2}}{\sqrt{2}}}{\sqrt{P_1} + \sqrt{P_2} + \sqrt{2Q_1}}, \tag{12.48}$$

$$\tilde{Q}_1 = \frac{c_1 + c_2}{2c_1} = \frac{\sqrt{P_1} + \sqrt{P_2} + \sqrt{2Q_1}}{2[\sqrt{P_1} + \frac{\sqrt{Q_1}+\sqrt{Q_2}}{\sqrt{2}}]}, \tag{12.49}$$

and $\tilde{P}_2 = 1 - \tilde{P}_1$, $\tilde{Q}_2 = 1 - \tilde{Q}_1$. Thus,

$$\langle\psi|D|\psi\rangle \sim \Big((c_1 + c_2)[u_1\tilde{P}_1 + u_2\tilde{P}_2] - 2c_1[v_1\tilde{Q}_1 + v_2\tilde{Q}_2]\Big) \geq 0. \tag{12.50}$$

However, as was emphasized in section 12.8, generally the quantities $\tilde{P}_1$, $\tilde{P}_2$ and $\tilde{Q}_1$, $\tilde{Q}_2$ cannot be interpreted as probabilities since $\tilde{P}_2$ and $\tilde{Q}_2$ can become negative (and, hence, $\tilde{P}_1$ and $\tilde{Q}_1$ can become larger than 1). Therefore, to keep to the probabilistic reasoning, we have to split the process of comparison of the lotteries in the four components (see section 12.7); each of this component can be interpreted as a comparison of two subjective utilities.

12.10 Mathematical Calculations

We start with calculation of average $\mathrm{tr} D_{B\to A}\sigma_A$.

$$\begin{aligned}
\mathrm{tr} D_{B\to A}\sigma_A &= \mathrm{tr}\left(\sum_{n,m} u_{nm} e^{i\gamma_{m_b\to n_a}} |n_a\rangle\langle m_b|\right) \\
&\quad \times \left(\sum_{i,j} \sqrt{P_i P_j} e^{i(\theta_{ai}-\theta_{aj})} |i_a\rangle\langle j_a|\right) \\
&= \sum_{i,j,n,m} u_{nm}\sqrt{P_i P_j} e^{i(\theta_{ai}-\theta_{aj}+\gamma_{m_b\to n_a})} \langle j_a|n_a\rangle\langle m_b|i_a\rangle
\end{aligned}$$

By using the definition of quantum transition (conditional) probabilities (see (12.4) and (12.5)), we obtain

$$\mathrm{tr} D_{B\to A}\sigma_A = \sum_{i,j,m} u_{jm}\sqrt{p(m_b|i_a) P_i P_j} e^{i(\theta_{i_a\to m_b}+\gamma_{m_b\to j_a}+\theta_{ai}-\theta_{aj})}$$

In the same way we obtain

$$\begin{aligned}
\mathrm{tr} D_{A\to B}\sigma_A &= -\mathrm{tr}\Big(\sum_{n,m} u_{nm} e^{i\gamma_{n_a\to m_b}} |m_b\rangle\langle n_a|\Big) \\
&\quad \times \Big(\sum_{i,j} \sqrt{P_i P_j} e^{i(\theta_{ai}-\theta_{aj})} |i_a\rangle\langle j_a|\Big) \\
&= -\sum_{i,j,n,m} u_{nm}\sqrt{P_i P_j} e^{i(\theta_{ai}-\theta_{aj}+\gamma_{n_a\to m_b})} \langle n_a|i_a\rangle\langle j_a|m_b\rangle \\
&= -\sum_{i,j,m} u_{im}\sqrt{p(j_a|m_b) P_i P_j} e^{i(\theta_{m_b\to i_a}+\gamma_{i_a\to m_b}+\theta_{ai}-\theta_{aj})}
\end{aligned}$$

This gives us expression (12.23) for $\mathrm{tr} D\sigma_A$. The calculations leading to expression (12.24) for $\mathrm{tr} D\sigma_B$ just repeat the previous ones.

Now we find traces for comparison operators and "transition states":

$$\begin{aligned}
\mathrm{tr} D_{B\to A}\sigma_{B\to A} &= \mathrm{tr}\Big(\sum_{n,m} u_{nm} e^{i\gamma_{m_b\to n_a}} |n_a\rangle\langle m_b|\Big) \\
&\quad \times \Big(\sum_{i,j} \sqrt{P_i Q_j} e^{i(\theta_{ai}-\theta_{bj})} |i_a\rangle\langle j_b|\Big) \\
&\quad \sum_{i,j,n,m} u_{nm}\sqrt{P_i Q_j} e^{i(\theta_{ai}-\theta_{bj}+\gamma_{n_a\to m_b})} \langle j_b|n_a\rangle\langle m_b|i_a\rangle \\
&= \sum_{i,j,n,m} u_{nm}\sqrt{p(n_a|j_b)p(m_b|i_a) P_i Q_j} e^{i(\theta_{i_a\to m_b}-\theta_{j_b\to n_a}+\gamma_{n_a\to m_b}+\theta_{ai}-\theta_{bj})}.
\end{aligned}$$

$$\begin{aligned}
\mathrm{tr} D_{A\to B}\sigma_{B\to A} &= -\mathrm{tr}\Big(\sum_{n,m} u_{nm} e^{i\gamma_{n_a\to m_b}} |m_b\rangle\langle n_a|\Big) \\
&\quad \times \Big(\sum_{i,j} \sqrt{P_i Q_j} e^{i(\theta_{ai}-\theta_{bj})} |i_a\rangle\langle j_b|\Big) \\
&= -\sum_{i,j,n,m} u_{nm}\sqrt{P_i Q_j} e^{i(\theta_{ai}-\theta_{bj}+\gamma_{n_a\to m_b})} \langle j_b|m_b\rangle\langle n_a|i_a\rangle \\
&= -\sum_{i,j} u_{ij}\sqrt{P_i Q_j} e^{i(\theta_{ai}-\theta_{bj}+\gamma_{i_a\to j_b})}.
\end{aligned}$$

In the same way

$$\begin{aligned}
&\mathrm{tr} D_{A\to B}\sigma_{A\to B} \\
&= -\sum_{i,j,n,m} u_{nm}\sqrt{p(m_b|i_a)p(n_a|j_b) P_i Q_j} e^{i(\theta_{j_b\to n_a}-\theta_{i_a\to m_b}+\gamma_{m_b\to n_a}+\theta_{bj}-\theta_{ai})}
\end{aligned}$$

and

$$\mathrm{tr} D_{B\to A}\sigma_{A\to B} = \mathrm{tr}\Big(\sum_{n,m} u_{nm} e^{i\gamma_{m_b\to n_a}} |n_a\rangle\langle m_b|\Big)$$
$$\times\Big(\sum_{i,j}\sqrt{P_i Q_j} e^{i(\theta_{bj}-\theta_{ai})}|j_b\rangle\langle i_a|\Big)$$
$$= \sum_{i,j} u_{ij}\sqrt{P_i Q_j} e^{i(\theta_{bj}-\theta_{ai}+\gamma_{j_b\to i_a})}.$$

Thus,

$$\Delta_1 = \frac{1}{2}\mathrm{tr}(D_{B\to A}\sigma_{A\to B} - D_{A\to B}\sigma_{B\to A})$$
$$= \sum_{i,j} u_{ij}\sqrt{P_i Q_j}\cos(\theta_{bj} - \theta_{ai} + \gamma_{j_b\to i_a}). \qquad (12.51)$$

Then,

$$\Delta_2 = \frac{1}{2}\mathrm{tr}(D_{B\to A}\sigma_{B\to A} - D_{A\to B}\sigma_{A\to B})$$
$$= \frac{1}{2}\sum_{i,j,n,m} u_{nm}\sqrt{p(n_a|j_b)p(m_b|i_a)P_i Q_j} e^{i(\theta_{i_a\to m_b}-\theta_{j_b\to n_a}+\gamma_{n_a\to m_b}+\theta_{ai}-\theta_{bj})}$$
$$-\frac{1}{2}\sum_{i,j,n,m} u_{nm}\sqrt{p(m_b|i_a)p(n_a|j_b)P_i Q_j} e^{i(\theta_{j_b\to n_a}-\theta_{i_a\to m_b}-\gamma_{n_a\to m_b}+\theta_{bj}-\theta_{ai})}.$$

Finally, we get

$$\Delta_2 = \sum_{i,j,n,m} u_{nm}\sqrt{p(m_b|i_a)p(n_a|j_b)P_i Q_j}$$
$$\times\cos(\theta_{j_b\to n_a} - \theta_{i_a\to m_b} + \gamma_{n_a\to m_b} + \theta_{bj} - \theta_{ai}). \qquad (12.52)$$

12.11 Concluding Remarks

We presented a quantum-like model of lottery selection based on representation of beliefs of an agent by pure quantum states. Subjective probabilities are mathematically realized in the framework of QP. Utility functions are borrowed from classical theory, but they are represented not only by their values. In the quantum representation, each event of realization of an outcome of a lottery and its utility are "blurred." Heuristically, one can say that each value $u_i = u(x_i)$ is surrounded by a cloud of information related to the event (A, x_i).

An agent process this information by using the rules of quantum information and QP. In general, this process is very complex; it combines fluctuations between preferences for different outcomes of lotteries. These fluctuations induce interference-type effects (constructive or destructive). The decision process which is formally represented by the comparison operator can be decomposed into a few subprocesses. Each of them can be formally treated as comparison of SEUs and interference factors (the latter express, in particular, risks related to lottery selection).

Model's structure is very rich, and further analysis of this structure will demand essential mathematical and interpretational efforts. We plan to continue research in this direction.

References

1. Adenier, G., Khrennikov, A. and Nieuwenhuizen, Th. M. (eds.) (2006). *Quantum Theory: Reconsideration of Foundations-3*, American Institute of Physics, Ser. Conference Proceedings 810, Melville, NY.
2. Adenier, G. and Khrennikov, A. (2006). Anomalies in EPR-Bell experiments. In: Adenier, G., Khrennikov, A. Yu. and Nieuwenhuizen, Th. M. (eds.) *Quantum Theory: Reconsideration of Foundations-3*, pp. 283–293. American Institute of Physics, Ser. Conference Proceedings 810, Melville, NY.
3. Adenier, G., Fuchs, C. and Khrennikov, A. (eds.) (2007). *Foundations of Probability and Physics-3*, American Institute of Physics, Ser. Conference Proceedings 889, Melville, NY.
4. Adenier, G. and Khrennikov, A. Yu. (2007). Is the fair sampling assumption supported by EPR experiments? *J. Phys. B*, 40 (1), 131–141.
5. Adenier, G., Khrennikov, A. Yu., Lahti, P., Manko, V. I., and Nieuwenhuizen, Th. M. (eds.) (2008). *Quantum Theory: Reconsideration of Foundations-4*, American Institute of Physics, Ser. Conference Proceedings 962, Melville, NY.
6. Aerts, D., Sozzo, S. and Tapia, J. (2012). *A Quantum Model for the Ellsberg and Machina Paradoxes*, Quantum Interaction, Lecture Notes in Computer Science, 7620, pp. 48–59.
7. Allais, M. (1953). Le comportement de l'homme rationnel devant le risque: critique des postulats et axiomes de l' cole amricaine, *Econometrica*, 21, 503–536.
8. Asano, M., Khrennikov, A., Ohya, M., Tanaka, Y. and Yamato, I. (2014). Violation of contextual generalization of the Leggett-Garg inequality for recognition of ambiguous figures, *Phys. Scr.*, T163, 014006.
9. Asano, M., Khrennikov, A., Ohya, M., Tanaka, Y. and Yamato, I. (2015). *Quantum Adaptivity in Biology: From Genetics to Cognition*, Springer: Heidelberg-Berlin-New York.

10. Asano, M., Basieva, I., Khrennikov, A., Ohya, M., Tanaka, Y. and Yamato, I. (2015). Quantum information biology: from information interpretation of quantum mechanics to applications in molecular biology and cognitive psychology, *Found. Phys.*, 45 (10), 1362–1378.
11. Asano, M., Basieva, I., Khrennikov, A., Ohya, M., and Tanaka, Y. (2017). A quantum-like model of selection behavior, *J. Math. Psychol.*, 78, 2–12.
12. Atmanspacher, H. and Fuchs, C. A. (eds.) (2014). *The Pauli-Jung Conjecture: And its Impact Today*, Imprint Academic.
13. Avis, D., Fischer, P., Hilbert, A. and Khrennikov, A. (2009). Single, complete, probability spaces consistent with EPR-Bohm-Bell experimental data. In: *Foundations of Probability and Physics-5*, AIP Conference Proceedings, 1101, 294–301.
14. Bagarello, F. (2012). *Quantum Dynamics for Classical Systems: With Applications of the Number Operator*; Wiley Ed.: New York, USA.
15. Bagarello, F. and Oliveri, F. (2013). A phenomenological operator description of interactions between populations with applications to migration, *Math. Models Methods Appl. Sci.*, 23 (03), 471–492.
16. Bagarello, F. and Haven, E. (2015). Towards a formalization of a two traders market with information exchange, *Phys. Scr.*, 90 (1), 015203.
17. Ballentine, L. E. (1989). The statistical interpretation of quantum mechanics, *Rev. Mod. Phys.*, 42, 358–381.
18. Ballentine, L. E. (1998). *Quantum Mechanics: A Modern Development*; WSP: Singapore.
19. Ballentine, L. E. (2001). Interpretations of probability and quantum theory. In: Khrennikov, A. Yu. (ed.) *Foundations of Probability and Physics, Quantum Probability and White Noise Analysis*, 13, 71–84, WSP: Singapore.
20. Barbera, P., Jost, J. T., Nagler, J., Tucker, J. A. and Bonneau, R. (2015). Tweeting from left to right: is online political communication more than an echo chamber?, *Psychol. Sci.*, 26 (10), 1531–1542.
21. Barlow, T. M., Bennett, R. and Beige, A. (2015). A master equation for a two-sided optical cavity, *J. Mod. Opt.*, 62 (Suppl 2), S11–S20.
22. Barros, J., Toffano, Z., Meguebli, Y. and Doan, B.-L. (2013). Contextual query using bell tests. In: *Proceedings of the Quantum Interaction*, Leicester, UK; Atmanspacher, H., Haven, E., Kitto, K. and Raine, D. (eds.); Springer: Berlin, Heidelberg, ISBN 978-3-642-54942-7.
23. Basieva, I., Pothos, E., Trueblood, J., Khrennikov, A. and Busemeyer, J. (2017). Quantum probability updating from zero prior (by-passing Cromwell's rule), *J. Math. Psychol.*, 77, 58–69.

24. Basieva, I., Khrennikova, P., Pothos, E. M., Asano, M. and Khrennikov, A. (2018). Quantum-like model of subjective expected utility, *J. Math. Econ.*, 78, 150–162.
25. Basti, G., Capolupo, A. and Vitiello, G. (2017). Quantum field theory and coalgebraic logic in theoretical computer science, *Prog. Biophys. Mol. Biol.*, 130, A, 39–52.
26. Bernardo, J. M. and Smith, A. F. M. (1994). *Bayesian Theory*, Wiley: Chichester.
27. Bernays, E. (1923). *Crystallizing Public Opinion*, Boni and Liveright: New York.
28. Bernays, E. (1928). *Propaganda*, Horace Liveright: New York.
29. Bernroider, G. and Summhammer, J. (2012). Can quantum entanglement between ion transition states effect action potential initiation? *Cogn. Comput.*, 4, 29–37.
30. Bernroider, G. (2017). Neuroecology: modeling neural systems and environments, from the quantum to the classical level and the question of consciousness, *J. Adv. Neurosc. Res.*, 4, 1–9.
31. Birkhoff, G. and von Neumann, J. (1936). The logic of quantum mechanics, *Ann. Math.*, 37 (4), 823–843.
32. Bohr, N. (1935). Can quantum-mechanical description of physical reality be considered complete? *Phys. Rev.*, 48, 696–702.
33. Bohr, N. (1938). The causality problem in atomic physics. In: J. Faye and H. J. Folse (eds.), (1987). *The Philosophical Writings of Niels Bohr, Vol. 4: Causality and Complementarity, Supplementary Papers* (Ox Bow Press: Woodbridge, CT), pp. 94–121.
34. Bohr, N. (1987). *The Philosophical Writings of Niels Bohr*, 3 Vols. (Ox Bow Press: Woodbridge, CT).
35. Bouchet, N. (2016). Russia's "militarization" of colour revolutions, *Policy Perspect.*, 4/2, 1–5.
36. Boyer-Kassem, T., Duchene, S. and Guerci, E. (2015). Quantum-like models cannot account for the conjunction fallacy, *Theory Decis.*, 10, 1–32.
37. Brandenburger, A. (2010). The relationship between quantum and classical correlation in games, *Games Econ. Behav.*, 69, 175–183.
38. Broekaert, J., Basieva, I., Blasiak, P. and Pothos, E. M. (2017). Quantum dynamics applied to cognition: a consideration of available options, *Philos. Trans. R. Soc. A*, 375 (2106), 375–405.

39. Brukner, C. and Zeilinger, A. (1999). Malus' law and quantum information, *Acta Phys. Slovaca*, 49 (4), pp. 647–652.
40. Brukner, C. and Zeilinger, A. (1999). Operationally invariant information in quantum mechanics, *Phys. Rev. Lett.*, 83 (17), pp. 3354–3357.
41. Brukner, C. and Zeilinger, A. (2009). Information invariance and quantum probabilities, *Found. Phys.*, 39, 677.
42. Brukner, C. (2015). *On the Quantum Measurement Problem*, Preprint arXiv:1507.05255 [quant-ph]; To be published in Proc. Conf., Quantum UnSpeakables II: 50 Years of Bell's Theorem (Vienna, 19–22 June 2014).
43. Brukner, C., Zukowski, M. and Zeilinger, A. (2001). *The Essence of Entanglement*, arXiv:quant-ph/0106119.
44. Busemeyer, J. R., Wang, Z. and Townsend, J. T., (2006). Quantum dynamics of human decision making, *J. Math. Psychol.*, 50, 220–241.
45. Busemeyer, J. R. and Bruza, P. D. (2012). *Quantum Models of Cognition and Decision*, Cambridge University Press: Cambridge.
46. Busemeyer, J. R., Pothos, E. M., Franco, R. and Trueblood, J. (2011). A quantum theoretical explanation for probability judgment errors, *Psychol. Rev.*, 118, 193–218.
47. Butora, M. (2007). OK 98: A campaign of slovak NGOS for free and fair elections. In: *Reclaiming Democracy: Civil Society and Electoral Change in Central and Eastern Europe*. Demes, P. and Forbrig, J. (eds.), German Marshall Fund, pp. 21–52.
48. Caves, C. M., Fuchs, C. A. and Schack, R. (2002). Quantum probabilities as Bayesian probabilities, *Phys. Rev. A*, 65, 022305.
49. Caves, C. M., Fuchs, C. A. and Schack, R. (2007). Subjective probability and quantum certainty, *Stud. Hist. Philos. Mod. Phys.*, 38, 255.
50. Chiribella, G., D'Ariano, G. M. and Perinotti, P. (2010). Probabilistic theories with purification, *Phys. Rev. A*, 81, 062348.
51. Chiribella, G., D'Ariano, G. M. and Perinotti, P. (2012). Informational axioms for quantum theory. In: *Foundations of Probability and Physics-6*, AIP Conf. Proc. 1424, pp. 270–279.
52. Choustova, O. (2007). Quantum Bohmian model for financial market, *Physica A*, 374 (1), 304–314.
53. Choustova, O. (2008). Application of Bohmian mechanics to dynamics of prices of shares: Stochastic model of Bohm–Vigier from properties of price trajectories, *Int. J. Theor. Phys.* 47 (1), 252–260.

54. Choustova, O. (2009). Quantum probability and financial market, *Inf. Sci.*, 179 (5), 478–484.

55. Collins, R. (2004). *Interaction Ritual Chains*, Princeton University Press: Princeton.

56. Colman, A. (2003). *Oxford Dictionary of Psychology*, Oxford University Press: New York, USA, p. 77.

57. Conte, E., Khrennikov, A., Todarello, O. and Federici, A. (2008). A preliminary experimental verification on the possibility of Bell inequality violation in mental states, *Neuroquantology*, 6 (3), 214–221.

58. D'Ariano, G. M. (2007). Operational axioms for quantum mechanics. In: Adenier et al. (eds.), *Foundations of Probability and Physics-3* AIP Conf. Proc. 889, pp. 79–105.

59. D'Ariano, G. M. (2011). Physics as information processing. In: *Advances in Quantum Theory*, AIP Conf. Proc. 1327, pp. 7–16.

60. D'Ariano, G. M. and Khrennikov, A. (2016). Preface of the special issue quantum foundations: information approach, *Philos. Trans. R. Soc. A*, 374, 20150244.

61. de Barros, A. J. (2012). Quantum-like model of behavioral response computation using neural oscillators, *Biosystems* 110, 171–182.

62. De Finetti, B. (1931). *Probabilismo*, Logos, 14, p. 163; transl. (1989). *Probabilism*, Erkenntnis, 31, pp. 169–223.

63. De Finetti, B. (2008). *Philosophical Lectures on Probability* (Springer Verlag: Berlin-New York).

64. Demes, P. and Forbrig, J. (eds.) (2007). *Reclaiming Democracy: Civil Society and Electoral Change in Central and Eastern Europe*, German Marshall Fund.

65. Dirac, P. A. M. (1995). *The Principles of Quantum Mechanics*, Clarendon Press: Oxford.

66. Dodonov, V. V. and Man'ko, V. I. (1997). Positive distribution description for spin states, *Phys. Lett. A*, 229, 335–339.

67. Dubois, F. (2006). On fractaquantum hypothesis, *Res-Systemica*, 5, article 55.

68. Dubois, F. and Toffano, Z. (2016). Eigenlogic: a quantum view for multiple-valued and fuzzy systems. In: *Lecture Notes in Computer Science, Proceedings of the Quantum Interaction*, San Francisco, CA, USA, July 20–22, 2016; de Barros, J., Coecke, B. and Pothos, E. (eds.); Springer, Cham; Vol. 10106, ISBN 978-3-319-52288-3.

69. Dubois, F. and Toffano, Z. (2017). Eigenlogic: a quantum view for multiple-valued and fuzzy systems. In: *Quantum Interaction*, QI 2016. LNCS, Vol. 10106. Springer, pp. 239–251.
70. Dzhafarov, E. N. and Kujala, J. V. (2012). Selectivity in probabilistic causality: where psychology runs into quantum physics, *J. Math. Psychol.*, 56, 54–63.
71. Dzhafarov, E. N., Zhang, R. and Kujala, J. V. (2015). Is there contextuality in behavioral and social systems? *Philos. Trans. R. Soc. A*, 374, 20150099.
72. Dzhafarov, E. N., Kujala, J. V., Cervantes, V. H., Zhang, R. and Jones, M. (2016). On contextuality in behavioral data, *Philos. Trans. R. Soc. A*, 374, 20150234.
73. Dzhafarov, E. N. and Kon, M. (2018). On universality of classical probability with contextually labeled random variables, *J. Math. Psychol.*, 85, 17–24.
74. Einstein, A., Podolsky, B. and Rosen, N. (1935). Can quantum-mechanical description of physical reality be considered complete? *Phys. Rev.*, 47 (10), 777–780.
75. Einstein, A. and Infeld, L. (1938). *The Evolution of Physics: The Growth of Ideas from Early Concepts to Relativity and Quanta*, Cambridge University Press: Cambridge.
76. Ellsberg, D. (1961). Risk, ambiguity, and the savage axioms, *Q. J. Econ.*, 75 (4), 643–669.
77. Erneux, T. and Glorieux, P. (2010). *Laser Dynamics*, Cambridge University Press.
78. Feynman, R. P. (1951). *The Concept of Probability in Quantum Mechanics*; Berkeley Symp. on Math. Statist. and Prob., Proc. Second Berkeley Symp. on Math. Statist. and Prob. (Univ. of Calif. Press), 533–541.
79. Feynman, R. and Hibbs, A. (1965). *Quantum Mechanics and Path Integrals* (McGraw-Hill: New York).
80. Fisher, S. and Bijelic, B. (2007). GLAS 99: Civil society preparing the ground for post-Tudman Croatia. In: Demes, P. and Forbrig, J. (eds.), *Reclaiming Democracy: Civil Society and Electoral Change in Central and Eastern Europe*, German Marshall Fund, pp. 53–78.
81. Fishwick, C. (2016). Why did people vote for Donald Trump? Voters explain, *The Gardian*, November 9.

82. Fleck, J. A. Jr. (1966). Quantum theory of laser radiation. I. Many-atom effects, *Phys. Rev.*, 149, 309–319.
83. Fox, M. (2006). *Quantum Optics An Introduction*, Oxford University Press.
84. Freud, S. (1923). *The Ego and the Id*, W.W. Norton and Company: New York, USA.
85. Freud, S. (1900). *The Interpretation of Dreams*. In Standard edition 1954–1974, Hogarth Press: London, UK, Vols. 4–5, pp. 1–627.
86. Freud, S. (1957). *The Standard Edition of Complete Psychological Works of Sigmund Freud*, Edited and Translated by J. Strachey, Vols. I–XXIV, The Hogarth Press: London.
87. Friederich, S. (2012). In defense of non-ontic accounts of quantum states, *Stud. Hist. Philos. Sci. B*, 44, 77–92.
88. Friederich, S. (2014). *Interpreting Quantum Theory: A Therapeutic Approach (New Directions in the Philosophy of Science)*, Palgrave Macmillan.
89. Fuchs, C. A. (2002). Quantum mechanics as quantum information (and only a little more). In: A. Khrennikov (ed.), *Quantum Theory: Reconsideration of Foundations*, Ser. Math. Model. 2 (Växjö University Press, Växjö), pp. 463–543.
90. Fuchs, C. A. (2002). *The anti-Växjö Interpretation of Quantum Mechanics, Quantum Theory: Reconsideration of Foundations*, Ser. Math. Model. 2 (Växjö University Press, Växjö), pp. 99–116.
91. Fuchs, C. (2007). Delirium quantum (or, where I will take quantum mechanics if it will let me). In: Adenier, G., Fuchs, C. and Khrennikov, A. Yu. (eds.), *Foundations of Probability and Physics-3*, Ser. Conference Proceedings, 889 (American Institute of Physics, Melville, NY), pp. 438–462.
92. Fuchs, C. A. and Schack, R. (2011). A quantum-Bayesian route to quantum-state space, *Found. Phys.* 41, 345.
93. Fuchs, C. A. (2012). *QBism, the perimeter of Quantum Bayesianism*, arXiv:1003.5209 [quant-ph].
94. Fuchs, C. A. and Schack, R. (2012). Bayesian conditioning, the reflection principle, and quantum decoherence. In: *Probability in Physics*, ser. The Frontiers Collection, Springer: Berlin-Heidelberg, pp. 233–247.
95. Fuchs, C. A. and Schack, R. (2013). Quantum-Bayesian coherence, *Rev. Mod. Phys.*, 85, 1693.

96. Fuchs, C. A. and Schack, R. (2014). QBism and the Greeks: why a quantum state does not represent an element of physical reality, *Phys. Scr.*, 90, 015104.

97. Fuchs, C. A., Mermin, N. D. and Schack, R. (2014). An introduction to QBism with an application to the locality of quantum mechanics, *Am. J. Phys.* 82, 749.

98. Fuchs, C. A. (2014). *My Struggles with the Block Universe*, http://arxiv.org/abs/1405.2390.

99. F. Fukuyama, (2013). The middle-class revolution, *The Wall Street Journal*, June 28.

100. Gonzales, R. and Wu, G. (1999). On the shape of the probability weighting function, *Cogn. Psych.* 38, 129–166.

101. Grauwin, S., Bertin, E., Lemoy, R. and Jensen, P. (2009). Competition between collective and individual dynamics, *PNAS*, 106 (49), 20622–20626.

102. Greenberger, D. M., Erez, N., Scully, M. O., Svidzinsky, A. A. and Zubairy, M. S. (2007). Planck, photon statistics and Bose–Einstein condensation, *Prog. Opt.* 50, 273–330.

103. Hameroff, S. (1994). Quantum coherence in microtubules. A neural basis for emergent consciousness? *J. Cons. Stud.*, 1, 91–118.

104. Haven, E. (2005). Pilot-wave theory and financial option pricing, *Int. J. Theor. Phys.* 44 (11), 1957–1962.

105. Haven, E. and Khrennikov, A. (2009). Quantum mechanics and violation of the sure-thing principle: the use of probability interference and other concepts, *J. Math. Psychol.*, 53, 378–388.

106. Haven, E. and Khrennikov, A. (2013). *Quantum Social Science*; Cambridge University Press.

107. Haven, E., Khrennikov, A. and Robinson, T. R. (2017). *Quantum Methods in Social Science: A First Course*; WSP: Singapore.

108. Haken, H. (1977). *Synergetics*, Springer.

109. Haken, H. (1981). *The Science of Structure: Synergetics*, Van Nostrand Reinhold: New York.

110. Haken, H. (1983). *Synergetics: An Introduction–Nonequilibrium Phase Transitions and Self-Organisation in Physics, Chemistry and Biology*, Springer.

111. Haken, H. (1985). *Laser Light Dynamics*, Vols. 1–2. North-Holland Amsterdam.

112. Hanbury Brown, R. and Twiss, R. Q. (1956). Correlation between photons in two coherent beams of light, *Nature*, 177, 27–29.

113. Heisenberg, W. (1930). *Physical Principles of Quantum Theory*, Chicago Univ. Press.

114. Hong, C. K., Ou, Z. Y. and Mandel, L. (1987). Measurement of subpicosecond time intervals between two photons by interference, *Phys. Rev. Lett.* 59 (18), 2044–2046.

115. Iurato, G. and Khrennikov, A. Yu. (2015). Hysteresis model of unconscious-conscious interconnection: exploring dynamics on m-adic trees, *p-Adic Numbers, Ultrametric Analysis, and Applications* 7 (4), 312–321.

116. Iurato, G., Khrennikov, A. and Murtagh, F. (2016). Formal foundations for the origins of human consciousness, *p-Adic Numbers, Ultrametric Analysis, and Applications*, 8 (4), 249–279.

117. Jaeger, C., Horn, G., Lux, T., Fricke, D., Frurst, S., Lass, W., Lin, L., Mandel, A., Meissner, F., Schreiber, S., Vesper, D. and Zwiener, R. (2009). *From the Financial Crisis to Sustainability? Potsdam, European Climate Forum*, Available online: www.european-climate-forum.net/index.php?id=ecfreports.

118. James, W. (1884). The dilemma of determinism, in (2009). In: *The Will to Believe and Other Essays in Popular Philosophy* (The Project Gutenberg Ebook), www.gutenberg.net, pp. 145–183.

119. James, W. (1882). On some hegelisms, in (2009). In: *The Will to Believe and Other Essays in Popular Philosophy* (The Project Gutenberg Ebook), www.gutenberg.net, pp. 263–298.

120. James, W. (1890). *The Principles of Psychology*, Henry Holt and Co.: New York, Reprinted 1983, Harvard Univ. Press: Boston.

121. Jung, C. G. (2001). *On the Nature of the Psyche*, Routledge Classics.

122. Jung, C. G. and Pauli, W. (2014). *Atom and Archetype: The Pauli/Jung Letters 1932–1958*, Princeton Univ. Press: Princeton.

123. Kahneman, D. and Tversky, A. (1972). Subjective probability: A judgment of representativeness, *Cogn. Psych.*, 3 (3), 430–454.

124. Kahneman, D. and Tversky, A. (1979). Prospect theory: an analysis of decision under risk, *Econometrica*, 47, 263–291.

125. Kahneman, D. and Tversky, A. (1984). Choices, values and frames, *Am. Psych.*, 39 (4), 341–350.

126. Kahneman, D. (2003). Maps of bounded rationality: psychology for behavioral economics, *Am. Econ. Rev.*, 93 (5), 1449–1475.

127. Kahneman, D. and Thaler., R. (2006). Utility maximization and experienced utility, *J. Econ. Perspect.*, 20, 221–232.

128. Khrennikov, A. Yu. (1994). *P-adic Valued Distributions and their Applications to the Mathematical Physics*, Kluwer Academic Publ.: Dordreht.

129. Khrennikov, A. (1997). *Non-Archimedean Analysis: Quantum Paradoxes, Dynamical Systems and Biological Models*, Kluwer Academic Publ.: Dordreht.

130. Khrennikov, A. (1999). Classical and quantum mechanics on information spaces with applications to cognitive, psychological, social and anomalous phenomena, *Found. Phys.* 29, 1065–1098.

131. Khrennikov, A. Yu. (1999). *Interpretations of Probability* (VSP Int. Sc. Publishers: Utrecht/Tokyo).

132. Khrennikov, A. (ed) (2001). *Foundations of Probability and Physics.* Series PQ-QP: Quantum Probability and White Noise Analysis 13, WSP: Singapore.

133. Khrennikov, A. (2000). Non-Kolmogorov probability models and modified Bell's inequality, *J. Math. Phys.* 41, 1768–1777.

134. Khrennikov, A. Yu. (2001). Linear representations of probabilistic transformations induced by context transitions, *J. Phys. A*, 34, 9965–9981.

135. Khrennikov, A. Yu. (2001). Origin of quantum probabilities. In: Khrennikov, A. (ed.), *Foundations of Probability and Physics* (Växjö-2000, Sweden; WSP, Singapore), pp. 180–200.

136. Khrennikov, A. (2001). Contextualist viewpoint to Greenberger-Horne-Zeilinger paradox, *Phys. Lett. A*, 278, 307–314.

137. Khrennikov, A. (2002). Växjö interpretation of quantum mechanics. In: *Quantum Theory: Reconsideration of Foundations*, Ser. Math. Modelling (Växjö Univ. Press), Vol. 2, pp. 163–170; Preprint: arXiv:quant-ph/0202107.

138. Khrennikov, A. (ed.) (2002). *Quantum Theory: Reconsideration of Foundations*, Ser. Math. Modelling in Physics, Engineering, and Cognitive Science, Vol. 2, Vaxjo Univ. Press: Växjö.

139. Khrennikov, A. (ed.) (2003). *Foundations of Probability and Physics-2*, Ser. Math. Modelling 5. Växjö Univ. Press, Växjö.

140. Khrennikov, A. (2003). Quantum-like formalism for cognitive measurements, *Biosystems*, 70, 211–233.

141. Khrennikov, A. (2003). Contextual viewpoint to quantum stochastics, *J. Math. Phys.*, 44, 2471–2478.

142. Khrennikov, A. (2004). On quantum-like probabilistic structure of mental information, *Open Syst. Inf. Dyn.* 11 (3), 267–275.

143. Khrennikov, A. Yu. (ed.) (2004). *Quantum Theory: Reconsideration of Foundations-2*, Ser. Math. Modelling 10. Växjö Univ. Press, Växjö.

144. Khrennikov, A. (2004). Växjö interpretation-2003: Realism of contexts. In: *Quantum Theory: Reconsideration of Foundations, Ser. Math. Model.*, Vol. 10 (Växjö Univ. Press), pp. 323–338.

145. Khrennikov, A. (2004). *Information Dynamics in Cognitive, Psychological, Social, and Anomalous Phenomena*, Ser.: Fundamental Theories of Physics, Kluwer, Dordreht.

146. Khrennikov, A. (2005). Reconstruction of quantum theory on the basis of the formula of total probability. In: *Foundations of Probability and Physics-3*, AIP Conf. Proc., 750 (American Institute of Physics, Melville, NY), pp. 187–218.

147. Khrennikov, A. (ed.) (2005). *Foundations of Probability and Physics-3*, American Institute of Physics, Ser. Conference Proceedings 750, Melville, NY.

148. Khrennikov, A. (2005). Schrödinger dynamics as the Hilbert space projection of a realistic contextual probabilistic dynamics, *Europhys. Lett.*, 69, 678–684.

149. Khrennikov, A. (2005). The principle of supplementarity: a contextual probabilistic viewpoint to complementarity, the interference of probabilities, and the incompatibility of variables in quantum mechanics, *Found. Phys.*, 35 (10), 1655–1693.

150. Khrennikov, A. (2006). Quantum-like brain: interference of minds, *BioSystems*, 84, 225–241.

151. Khrennikov, A. (2008). Bell-Boole inequality: Nonlocality or probabilistic incompatibility of random variables? *Entropy*, 10, 19–32.

152. Khrennikov, A. (2009). *Interpretations of Probability*, 2nd edn., De Gruyter: Berlin.

153. Khrennikov, A. (2009). *Contextual Approach to Quantum Formalism*, Springer: Berlin-Heidelberg-New York.

154. Khrennikov, A. (2010). *Ubiquitous Quantum Structure: From Psychology to Finances*, Springer: Berlin-Heidelberg-New York.

155. Khrennikov, A. (ed.) (2010). *Quantum Theory: Reconsideration of Foundations-5*, American Institute of Physics, Ser. Conference Proceedings 1232, Melville, NY

156. Khrennikov, A. (2011). Quantum-like model of processing of information in the brain based on classical electromagnetic field, *Biosystems*, 105 (3), 250–262.

157. Khrennikov, A. (2011). Prequantum classical statistical field theory: Schrödinger dynamics of entangled systems as a classical stochastic process, *Found. Phys.*, 41, 317–329.

158. Khrennikov, A. (2012). Towards a field model of prequantum reality, *Found. Phys.*, 42, 725–741.

159. Khrennikov, A. (2012). Quantum probabilities and violation of CHSH-inequality from classical random signals and threshold type detection scheme, *Prog. Theor. Phys.*, 128, 31–58.

160. Khrennikov, A., Nilsson, B. and Nordebo, S. (2012). On an experimental test of prequantum theory of classical random fields: an estimate from above of the coefficient of second-order coherence, *Int. J. Quantum Inf.*, 10, 1241014.

161. Khrennikov, A. and Basieva, I. (2014). Quantum model for psychological measurements: from the projection postulate to interference of mental observables represented as positive operator valued measures, *NeuroQuantology*, 12, 324–336.

162. Khrennikov, A. (2014). *Beyond Quantum*, Pan Stanford Publishing: Singapore.

163. Khrennikov, A. and Basieva, I. (2014). Possibility to agree on disagree from quantum information and decision making, *J. Math. Psychol.*, 62 (3), 1–5.

164. Khrennikov, A. (2015). Two-slit experiment: quantum and classical probabilities, *Phys. Scr.*, 90, 074017.

165. Khrennikov, A. (2015). CHSH inequality: quantum probabilities as classical conditional probabilities, *Found. Phys.*, 45, 711–725.

166. Khrennikov, A. (2015). Quantum-like modeling of cognition, *Front. Phys.*, https://doi.org/10.3389/fphy.2015.00077

167. Khrennikov, A. (2015). Quantum-like model of unconscious-conscious dynamics, *Front. Psychol.*, 6, 997.

168. Khrennikov, A. (2015). Towards information lasers, *Entropy*, 17 (10), 6969–6994.

169. Khrennikov, A. (2016). Social laser: action amplification by stimulated emission of social energy, *Phil. Trans. R. Soc.*, 374 (2054), 20150094.

170. Khrennikov, A. (2016). Analog of formula of total probability for quantum observables represented by positive operator valued measures, *Int. J. Theor. Phys.*, 55, 859–874.

171. Khrennikov, A. (2016). Quantum Bayesianism as the basis of general theory of decision-making, *Phil. Trans. R. Soc. A*, 374, 20150245.

172. Khrennikov, A. (2016). Bell could become the Copernicus of probability, *Open Syst. Inf. Dyn.*, 23, 1650008.

173. Khrennikov, A. (2016). *Probability and Randomness: Quantum Versus Classical*, Imperial College Press.

174. Khrennikov, A. (2017). Quantum epistemology from subquantum ontology: quantum mechanics from theory of classical random fields, *Ann. Phys.*, 377, 147–163.

175. Khrennikov, A. (2018). External observer reflections on QBism, its possible modifications, and novel applications. In: Khrennikov, A. and Toni, B. (eds.), *Quantum Foundations*, STEAM-H: Science, Technology, Engineering, Agriculture, Mathematics & Health, Springer: Cham, pp. 93–118.

176. Khrennikov, A. (2018). Social laser model: from color revolutions to Brexit and election of Donald Trump, *Kybernetes*, 47 (2), 273–278.

177. Khrennikov, A. (2018). External observer reflections on QBism, its possible modifications, and novel applications. In: Khrennikov, A. and Toni, B. (eds.), *Quantum Foundations*, Probability and Information, STEAM-H: Science, Technology, Engineering, Agriculture, Mathematics and Health, Springer: Cham, pp. 93–118.

178. Khrennikov, A. (2018). Towards better understanding QBism, *Found. Sci.*, 23 (1), 181–195.

179. Khrennikov, A. and Alodjants, A. (2018). Classical (local and contextual) probability model for Bohm-Bell type experiments: no-signaling as independence of random variables, *Entropy*, 21 (2), 157.

180. Khrennikov, A., Basieva, I., Pothos, E. M. and Yamato, I. (2018). Quantum probability in decision making from quantum information representation of neuronal states, *Sci. Rep.*, 8, 16225.

181. Khrennikov, A., Alodjants, A. Trofimova, A. and Tsarev, D. (2018). On interpretational questions for quantum-Like modeling of social lasing, *Entropy*, 20 (12), 921.

182. Khrennikov, A., Toffano, Z. and Dubois, F. (2019). Concept of information laser: from quantum theory to behavioural dynamics, *Eur. Phys. J. Spec. Top.*, 227 (15–16), 2133–2153.

183. Khrennikov, A. (2019). Classical versus quantum probability: Comments on the paper "On universality of classical probability with contextually labeled random variables" by E. Dzhafarov and M. Kon, *J. Math. Psych.* 89, 87–92.

184. Khrennikova, P., Haven, E. and Khrennikov, A. (2014). An application of the theory of open quantum systems to model the dynamics of party governance in the US Political System, *Int. J. Theor. Phys.*, 53 (4), 1346–1360.

185. Khrennikova, P., Haven, E. (2016). Instability of political preferences and the role of mass-media: a dynamical representation in a quantum framework, *Philos. Trans. R. Soc. A*, 374, 20150106.

186. Khrennikova, P. (2016). Quantum dynamical modeling of competition and cooperation between political parties: the coalition and non-coalition equilibrium model, *J. Math. Psychol.*, 71, 39–50.

187. Khrennikova, P. (2017). Modeling behavior of decision makers with the aid of algebra of qubit creation-annihilation operators, *J. Math. Psych.*, 78, 76–85.

188. Krastev, I. (2013). *In Mistrust we Trust: Can Democracy Survive When we don't Trust Our Leaders?*, TED Conferences Press: Monterey.

189. Krastev, I. (2014). *Democracy Disrupted: The Global Politics of Protest*, Penn Press: Philadelphia.

190. Kofler, J. and Zeilinger, A. (2010). Quantum information and randomness, *Eur. Rev.*, 18, 469–480.

191. Kolmolgoroff, A. N. (1933). *Grundbegriffe der Wahrscheinlichkeitsrechnung*, (Springer-Verlag, Berlin); Kolmolgorov, A. N. (1936). *The Basic Notions of Probability Theory*, FIAZ: Moscow.

192. Lax, M., Cai, W. and Xu, M. (2006). *Random Processes in Physics and Finance*, Oxford University Press.

193. Machina, M. J. (2009). Risk, ambiguity, and the dark-dependence axioms, *Am. Econ. Rev.*, 99 (1), 385–392.

194. Mackey, G. W. (1963). *Mathematical Foundations of Quantum Mechanics* (W. A. Benjamin Inc.: New York).

195. Mancini, S., Man'ko, V. I. and Tombesi, P. (1996). Symplectic tomography as classical approach to quantum systems, *Phys. Lett. A*, 213, 1–6.

196. Man'ko, V. I. and Man'ko, O. V. (1997). Spin state tomography, *J. Exp. Theor. Phys.*, 85, 430–434.

197. Man'ko, M. A. and Man'ko, V. I. (2018). New entropic inequalities and hidden correlations in quantum suprematism picture of qudit states, *Entropy*, 20, 692.

198. Mason, P. (2012). *Why It's Kicking Off Everywhere: The New Global Revolutions*, Verso: London.

199. Mermin, N. D. (2017). Why QBism is not the Copenhagen interpretation and what John Bell might have thought of it. In: Bertlmann, R. and Zeilinger, A. (eds.), *Quantum [Un]Speakables II. The Frontiers Collection,* Springer: Cham.
200. Mitchell, L. A. (2012). *The Color Revolutions*, University of Pennsylvania Press.
201. Ning, C. Z. (2013). What is laser threshold?, *IEEE J. Sel. Top. Quantum Electron.*, 19 (4), 6516933.
202. Orszag, A. (1968). *Les Lasers, Principes, Realizations, Applications*, chap. V, Masson et Cie: Paris.
203. Ozmon, H. A. and Craver, S. M. (2007). *Philosophical Foundations of Education*, Prentice Hall Publisher.
204. Penrose, R. (1989). *The Emperor's New Mind*, Oxford Univ. Press: New-York.
205. Plotnitsky, A. (2001). Reading Bohr: complementarity, epistemology, entanglement, and decoherence. In: Gonis, A. and Turchi, P. (eds.), *Proc. NATO Workshop "Decoherence and its Implications for Quantum Computations,"* IOS Press: Amsterdam, pp. 3–37.
206. Plotnitsky, A. (2006). *Reading Bohr: Physics and Philosophy*, Springer: Heidelberg-Berlin-New York.
207. Plotnitsky, A. (2009). *Epistemology and Probability: Bohr, Heisenberg, Schrödinger and the Nature of Quantum-Theoretical Thinking*, Springer: Berlin and New York.
208. Plotnitsky, A. (2012). *Niels Bohr and Complementarity: An Introduction*, Springer: Berlin and New York.
209. Plotnitsky, A. and Khrennikov, A. (2015). Reality without realism: on the ontological and epistemological architecture of quantum mechanics, *Found. Phys.*, 45 (10), 269–1300.
210. Pothos, E. and Busemeyer, J. R. (2009). A quantum probability explanation for violations of 'rational' decision theory, *Proc. R. Soc. B*, 276, 2171–2178.
211. Pothos, E. M. and Busemeyer, J. R. (2013). Can quantum probability provide a new direction for cognitive modeling?, *Behav. Brain Sci.*, 36, 255–274.
212. Priyashanka, K. M. and Wijewardena Gamalath, K. A. I. L. (2015). Coherent states in a laser cavity, *Int. Lett. Chem. Phys. Astron.*, 56, 47–62.

213. Putin, V. (2017). We must stop a Ukraine-style 'coloured revolution' in Russia, *The Telegraph*, January 1.

214. Ramsey, F. P. (1931). Truth and probability. In: Braithwaite, R. B. (ed.), *The Foundations of Mathematics and other Logical Essays*, Harcourt, Brace and Company: New York, p. 156.

215. Ricciardi, L. M. and Umezawa, H. (1967). Brain and physics of many-body problems, *Kibernetik*, 4 (2), 44–48.

216. Savage, L. J. (1954). *The Foundations of Statistics*, John Wiley & Sons: New York.

217. Schawlow, A. L. and Townes, C. H. (1958). Infrared and optical masers, *Phys. Rev.*, 112, 1940–1949.

218. Schmidt, E. and Cohen, J. (2013). *The New Digital Age: Reshaping the Future of People, Nations and Business*, Knopf: New York.

219. Schrödinger, E. (1935). Die gegenwurtige Situation in der Quantenmechanik, *Naturwissenschaften*, 23, 807–812; 823–828; 844–849.

220. Schrödinger, E. (1989). *Statistical Thermodynamics*, Dover Publications.

221. Sozzo, S. (2014). A quantum probability explanation in Fock space for borderline contradictions, *J. Math. Psychol.*, 58 (1), 1–12.

222. Timpson, S. G. (2008). Quantum Bayesianism: a study, *Stud. Hist. Philos. Sci. B*, 39, 579–609.

223. Timpson, S. G. (2013). *Quantum Information Theory and the Foundations of Quantum Mechanics* (Oxford Philosophical Monographs), Oxford University Press: Oxford.

224. Toffano, Z. (1997). Investigation of threshold transition in semiconductor lasers, *IEEE J. Sel. Top. Quantum Electron.*, 3 (2), 485–490.

225. Travers, T. (2016). Why did people vote for Brexit? Deep-seated grievances lie behind this vote, *British Politics and Policy*.

226. Tucker, J. A. (2007). Enough! Electoral fraud, collective action problems, and post-communist coloured revolutions, *Perspect. Politics*, 5 (3), 537–553.

227. Tversky, A. and Kahneman, D. (1974). Judgment under uncertainty: heuristics and biases, *Science*, 185 (4157), 1124–1131.

228. Tversky, D. and Kahneman, D., (1992). Advances in prospect theory: cumulative representation of uncertainty, *J. Risk Uncertainty*, 5, 297–323.

229. Umezawa, H. (1993). *Advanced Field Theory: Micro, Macro and Thermal Concepts*, AIP: New York.

230. Vitiello, G. (1995). Dissipation and memory capacity in the quantum brain model, *Int. J. Mod. Phys. B*, 9 (08), 973–989.

231. Vitiello, G. (2001). *My Double Unveiled: The Dissipative Quantum Model of Brain*, Advances in Consciousness Research, John Benjamins Publishing Company.

232. von Mises, R. (1919). Grundlagen der Wahrscheinlichkeitsrechnung, *Math. Z.*, 5, 52–99.

233. von Mises, R. (1957). *Probability, Statistics and Truth*, Macmillan: London.

234. von Mises, R. (1964). *The Mathematical Theory of Probability and Statistics*, Academic: London.

235. von Neuman, J. (1955). *Mathematical Foundations of Quantum Mechanics*, Princeton University Press: Princenton.

236. von Neumann, J. and Morgenstern, O. (1953). *Theory of Games and Economic Behavior*, Princeton University Press: Princeton.

237. Wigner, E. (1932). On the quantum correction for thermodynamic equilibrium, *Phys. Rev.* 40, 749–759.

238. Zeilinger, A. (1999). A foundational principle for quantum mechanics, *Found. Phys.*, 29 (4), 631–643.

Index